THE TANGLED MIND

Unraveling the Origin of Human Nature

THE TANGLED MIND

Unraveling the Origin of Human Nature

NICK KOLENDA

Copyright © 2019, 2020 Nick Kolenda
www.NickKolenda.com

Published by Kolenda Group LLC

ISBN: 978-1-7339789-0-3 (paperback)

All rights reserved. No part of this publication may be reproduced or distributed in any form or by any means, or stored in a database or retrieval system, without the prior written permission of the publisher.

Contents

Introduction vii

PART 1—ORIGIN 1
   1. Scaffolding 3
   2. Simulation 11

PART 2—PRIMITIVES 25
   3. Size 27
   4. Object 37
   5. Location 47
   6. Distance 59
   7. Motion 71
   8. Shape 87
   9. Orientation 103
   10. Sound 111
   11. Physiology 125
   12. Emotions 137
   13. Color 149
   14. People 163

PART 3—APPLICATIONS 181
   15. Beauty 183
   16. Morality 195

17. Religion 213
18. Literature 231
19. Time 251

Conclusion 263
References 277

Introduction

WHICH IS BRIGHTER: sneeze or cough?

Weird question to open a book, right? Well . . . did you guess a sneeze? Strangely, most people do.

But why did you choose a sneeze? Brightness and sneezing are separate domains, yet—somehow—they feel oddly similar, don't they? Why is that?

Turns out, this example isn't isolated; many sensory concepts are connected without your awareness. Even worse, these hidden connections influence your perception and behavior every day. Heck, even right now. This book has sensory components—size, shape, weight—which are influencing your perception of the content. Your perception of me. In this sentence. And this one too.

Or how about words and punctuation—like the dash in this sentence? Does that matter? You bet. Even CAPITAL letters? Yup, them too. Or how about length? These sentences are pretty short. What if I wrote a long sentence, perhaps separated by a semi-colon; would the mere length—an irrelevant characteristic—change your perception of this sentence, including the inherent meaning? It sure would.

In a nutshell, this book explains how sensory concepts influence your perception of the world. I'll break down the academic research in a concise and lighthearted way, and I'll explain the practical applications. Alongside every business application, I'll describe a philanthropic application that can help people. Improve lives. Save lives.

In order to disentangle this sensory influence, we need to retrace the origin of perception and knowledge. Today, you understand many abstract concepts—from capitalism to bromance. But how did it all happen? How did you learn any abstract concept?

Humans learn by association: You relate new concepts to existing knowledge. But you might notice a problem. In order to learn Concept Z, you attached the meaning to an earlier concept, like Concept Y. Well, how did you learn Concept Y? You did the same thing: You attached Concept Y to an earlier concept, perhaps Concept X. Hopefully you see a pattern.

Every concept is connected to an earlier concept. So then, what happens if we retrace those steps? Like an endless string of knowledge, what if we keep pulling? Wouldn't we find a starting point? And, if so, what would it be? Wouldn't these concepts be infused into everything that we know today?

I'm writing this book because my curiosity got the best of me: *I needed to pull that string.* For the past few years, I've been researching the origin of our perception to find the starting point. Once I discovered it, I felt an epiphany: suddenly everything made sense. With my newfound knowledge, I could take any concept—concrete or abstract—and I could see the traces of that starting point. Today, everything that you understand is infused with the initial concepts that you learned in this world. Your perception, right now—in this very sentence—has been sculpted by those primitive concepts.

Pulling that string changed my view of the world; it literally changed my life. In this book, we'll pull that string together.

1
Origin

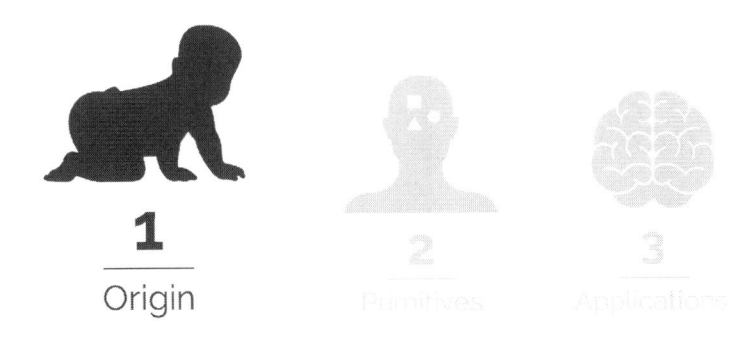

2
Primitives

3
Applications

In this part, you'll discover why sensory concepts distort your perception of the world. The answer can be found in the everyday lives of babies.

1

Scaffolding

YOU SEEM TO BE charging through this book. Let's reward you with an exercise.

> Your meeting next Wednesday has been moved forward two days. Quick. What day is the meeting now?

Got your answer?

You might have noticed that the term "forward" is ambiguous; the meeting could have been pushed toward you (Monday) or in front of you (Friday). However, you were more likely to choose Friday because of a hidden cue. Can you spot it?

The Beginning of Your Knowledge

Look around . . . you are surrounded by sensory concepts. Even if you close your eyes, you can't escape them while navigating the world.

Sensory concepts are so pervasive that they fuel everything you learn. Consider the abstract concept of time. You can't see time. You can't feel time. Nobody can teach a baby by saying: "Hey you, baby, this is time." I tried. It doesn't work.

So, how did you learn this abstract concept? You can spot a clue with language: You describe time with sensory concepts.

> We could spend a LONG time discussing this idea, but let's put these examples BEHIND us so that we can keep MOVING through this book.

You learned time by "scaffolding" this knowledge onto a sensory foundation (Williams, Huang, & Bargh, 2009). You took concepts that you already understood—size, orientation, motion—and you extended these ideas into the abstract territory of time. Thanks to this process, your concept of time was (and still is) infused with sensory concepts. Every time that you conceptualize time, you are painting this mental imagery with a sensory framework.

Remember the meeting on Wednesday? Why did you assume that it was pushed "forward" to Friday?

Immediately before posing that question, I mentioned that you were "charging" through this book. That subtle wording altered your mental imagery of time. You could have painted this imagery with two types of motion (see Lakoff & Johnson, 1999):

1. **Moving Time.** The meeting is moving toward you.
2. **Moving Observer.** You are moving toward the meeting.

A few simple words (e.g., "charging through this book") instilled a mental image in which *you* were moving through time. As a result, you believed that the meeting was pushed forward in *your* direction of motion (in this case, Friday). Researchers posed that question at an airport, and they noticed that arriving passengers were more likely to say Friday because of their recent bodily motion (Boroditsky & Ramscar, 2002).

Don't worry about the details for now. Focus on the overall takeaway: Your knowledge rests upon a sensory foundation. Sensory concepts are painting all of the mental images that you create.

Let's examine this process in more detail.

How You Learn New Concepts

You might be familiar with Ivan Pavlov who noticed that his dogs would salivate at the ring of a bell. Pavlov frequently rang a bell before feeding his dogs, and his dogs associated the bell with food.

Humans learn in a similar way. When two concepts frequently appear together, you connect them in your brain. Activating one concept will activate the other. It's called *spreading activation*.

The next few sections will illustrate this idea, and you'll see how this mechanism can bind two concepts that are vastly different.

UP and POWER

UP is a spatial location, while POWER is a subjective feeling. These concepts are inherently different, yet they are connected:

- When people see words about power, such as *king*, they look upward (Pecher, Van Dantzig, Boot, Zanolie, & Huber, 2010).
- In a visual hierarchy of employees, managers seem more powerful with longer vertical lines beneath them (Giessner & Schubert, 2007).
- Men seem more powerful (and attractive) when their pictures appear toward the top of a dating profile (Meier & Dionne, 2009).

UP and POWER are connected in your brain. Why? We can answer that question by becoming a baby . . . *fwoosh* . . . or whatever sound resembles becoming a baby.

Figure 1A. In order to understand adults, you need to place yourself in the mindset of a baby.

Alright, we're a baby. Look around you. Nothing makes sense, does it? We have no control, and we're at the whim of our parents. Oh look, they're coming now. They seem like giants, don't they? Uh oh. Wait. What are they doing? They're picking us up. Agh! Where are we going? What are we doing? We can't stop them.

Babies experience that sensation all the time. They don't understand it fully; they just feel a supreme force controlling them. Thus, a concept of POWER starts firing.

Meanwhile, they also notice that their parents are vertical giants. Therefore, two concepts—UP and POWER—are firing simultaneously in their brain. Both concepts are dispersing waves of activation that merge into a unitary circuit. In specific terms:

> When two neuronal groups, A and B, fire at the same time, activation spreads outward . . . When the activation spreading from A meets the activation spreading from B, a link is formed, and the link gets stronger the more A and B fire together (Lakoff, 2008, pp. 19–20).

Put simply: Two concepts become connected in your brain if they frequently occur simultaneously.

Let's see this idea with another connection to UP.

UP and MORE

Have you ever noticed that numbers move upward?
   Prices can INCREASE.
   Temperatures can RISE.
   Stocks can CLIMB.
   You might be thinking: *Well of course numbers rise . . . that's their nature.*
   Is it, though? Objects in the sensory world can move up and down, but numbers are intangible symbols with no inherent directionality:

> Elevators and airplanes can go up and down, literally. By contrast, the price of eggs, the rate of unemployment, the popularity of a politician, the value of the Yen, and the temperature of the air outside your window can only go "up" and "down" metaphorically (Casasanto, & Bottini, 2014b, p. 140).

So, why does it feel natural to conceptualize numbers to be moving upward? You can blame our sensory world.
   Place this book onto something else, and the pile will get higher. During this exposure, two concepts—UP and MORE—are activated in your brain, and this dual activation binds these concepts.
   Still confused? Let's break it down with a final example.

WARMTH and AFFECTION

Language is filled with hidden metaphors. For example, we describe AFFECTION in terms of WARMTH:

> I have a WARM personality, so I won't give you a COLD or ICY STARE. And I won't give you the COLD SHOULDER. That'd be COLD-BLOODED of me.

Why does this metaphor feel natural? As a baby, you experienced both concepts simultaneously:

> ... for an infant, the subjective experience of affection is typically correlated with the sensory experience of warmth, the warmth of being held ... the associations are automatically built up between the two domains. Later, during a period of differentiation, children are then able to separate out the domains, but the cross-domain associations persist (Lakoff & Johnson, 1999, p. 46).

Babies experience WARMTH and AFFECTION every time they are held, so they perceive these concepts to be identical. Eventually they disentangle these ideas, but the shared circuitry remains (Inagaki & Eisenberger, 2013).

Today, activating one concept will activate the other concept. If you activate WARMTH, you activate AFFECTION:

- ▶ **Touch.** While holding hot coffee, people judged others to be more affectionate (Williams & Bargh, 2008a).
- ▶ **Weather.** During warm weather, people bet on the "favorite" horse at a racetrack because they trust other people (Huang, Zhang, Hui, & Wyer Jr, 2014).

Likewise, activating AFFECTION will activate WARMTH:

- ▶ **Rejection.** People who are socially excluded feel physically colder (Zhong & Leonardelli, 2008). Even monkeys feel colder when excluded (McFarland, et al., 2015).
- ▶ **Acceptance.** When people dine with somebody else, the room seems warmer (after controlling for body heat; Lee, Rotman, & Perkins, 2014).

Perhaps most interesting, people who seek one concept will unknowingly seek the other concept. If you seek AFFECTION, you unknowingly seek WARMTH:

- ▶ **Bathing.** Lonely people take warmer baths and showers (Bargh, & Shalev, 2012).
- ▶ **Food.** Lonely people prefer warmer food and drinks, such as soup and coffee (Zhong & Leonardelli, 2008).

Or, if you are seeking WARMTH, you also seek AFFECTION:

- ▶ **Movies.** During cold weather, people spend more money on romantic movies (compared to actions, comedies, or thrillers; Hong & Sun, 2011).
- ▶ **Social Activities.** While completing a survey outside in winter, people showed stronger interest in social activities, like visiting their parents (vs. listening to a good lecture; Zhang, & Risen, 2014).

In this book, you'll discover an abundance of other metaphors in language. And it's not just English. These metaphors occur in all languages because humans live in the same sensory world:

> ... [metaphors] are learned by the hundreds the same way all over the world because people have the same bodies and basically the same relevant environments (Lakoff, 2008, p. 26).

In the next chapter, we'll continue unraveling this idea by examining a related nuance—*simulation*—in more detail.

Summary

You were thrust into a world with sensory concepts. In this book, you will encounter different verbiage—sensory elements, sensory traits, sensory ideas—but they're all the same.

You scaffold your knowledge onto this foundation, inserting these sensory ideas into abstract domains. Whenever you conceptualize an intangible concept, such as time, you are painting this mental imagery with a sensory framework.

You also connect many disparate concepts, such as WARMTH and AFFECTION. Today, activating one concept will activate other concepts that are connected to it.

2
Simulation

Figure 2A

LOOK AT THE PICTURE of me juggling.
Now, determine whether each word is depicted in that image:

- Catching
- Kicking
- Punching

Like most people, you probably noticed that catching was the only concept depicted. And that's correct. However, did you struggle more with punching than kicking? Punching contained an intrinsic quality that slowed your reaction time. Can you spot the reason?

Types of Simulation

Simulation might sound complicated, but you saw this idea in the previous chapter. Remember the meeting that was pushed forward? You understood this scenario by creating a mental picture. I'll refer to these mental pictures as "simulations."

In fact, you are simulating right now while reading these sentences. Compare these sentences:

- ▶ John put the pencil in the cup.
- ▶ John put the pencil in the drawer.

You understood those sentences by "simulating" each event. Researchers derived this conclusion through reaction times: When people read the first sentence with the pencil in a cup, they could spot a vertical pencil faster than a horizontal pencil (Stanfield & Zwaan, 2001).

In another study, participants read these sentences:

- ▶ The ranger saw the eagle in the sky.
- ▶ The ranger saw the eagle in the nest.

Afterward, researchers asked them to indicate whether an image (an eagle) was mentioned, and they answered this question faster if the eagle matched the implied orientation (see Figure 2B; Zwaan, Stanfield, & Yaxley, 2002).

Simulation   13

Figure 2B

Or consider vividness:

- ▶ Through clean goggles, the skier could identify the moose.
- ▶ Through fogged goggles, the skier could hardly identify the moose.

Participants who read "fogged goggles" were quicker to see a moose in a low-resolution image (Yaxley & Zwaan, 2007).

It even happens with abstract sentences:

- ▶ John opened the book, and an hour later, he finished it.
- ▶ John opened the book, and the next day, he finished it.

In the second sentence, people had more trouble remembering that John "opened" the book (Zwaan, 1996). Why? John finished the book in the more "distant" future. Participants needed to traverse backward across a farther distance to reach this event, which slowed their reaction time. That claim sounds radical, but hopefully it becomes less radical throughout this book.

The takeaway: You understand sentences by translating words into a mental picture.

Let's examine four types of mental pictures.

1. Concepts

Grab a pen and draw a teacup with your weaker hand. This example will serve multiple purposes.

Earlier, you read sentences that described the orientation of a pencil (e.g., pencil in cup). But what if you imagined a pencil without any descriptors? No orientation. No shape. No color. What would you imagine? In this case, you would imagine a stereotypical depiction of a pencil. I call it a *canonical prototype.*

Does your teacup look similar to Variation A in Figure 2C? Most people draw an eerily similar image (Palmer, 1981).

But why? You could have drawn many styles of teacups with different perspectives, features, and orientations (e.g., Variation B). Why did you draw Variation A?

Throughout your life, you see a plethora of teacups. You can't simulate all possible combinations, so your brain selects the traits that are most identifiable or common—a prototypical size, color, perspective, and more. Canonical prototypes will play a major role in this book, so we'll revisit this concept later.

Let's try another exercise. Think of a car.

Thinking of one?

Not only are you conceptualizing a prototypical car, but you are also activating a barrage of sensory experiences:

> On conceptualising CAR, for example, the visual system might become partially active as if a car were present.

Figure 2C

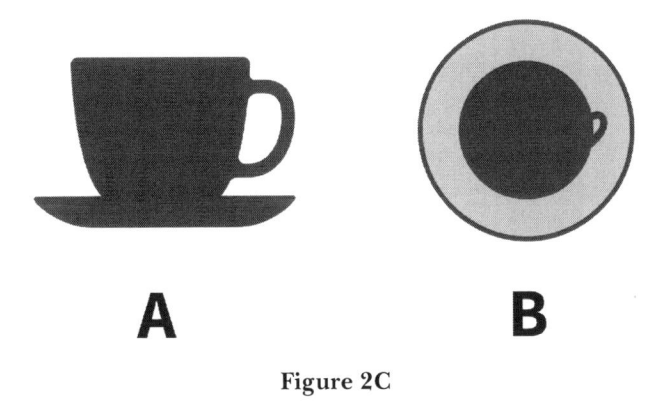

Similarly the auditory system might reenact states associated with hearing a car, the motor system might reenact states associated with driving a car, and the limbic system might reenact emotional states associated with enjoying the experience of driving (again all at the neural level; Barsalou, 2003, p. 523).

The next section will explore the motor and physiological aspects of simulation.

2. Actions

Look at the teacup in Variation A of Figure 2C. See the handle? Right-handed people prefer handles on the right, whereas left-handed people prefer handles on the left (Elder & Krishna, 2011). You prefer whichever orientation will help you imagine grabbing the handle. Interestingly, the effect disappears when people hold something (e.g., tennis ball) because neither group can simulate the interaction (Shen & Sengupta, 2012).

Many products, such as detergent, strategically place the handle on the right so that right-handed people, the majority of the population, can simulate the interaction more vividly (see Figure 2D).

Handle on Left   Handle on Right

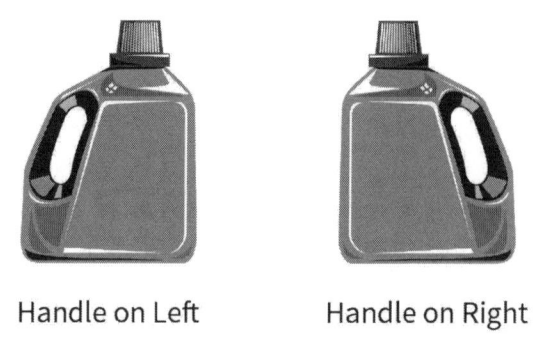

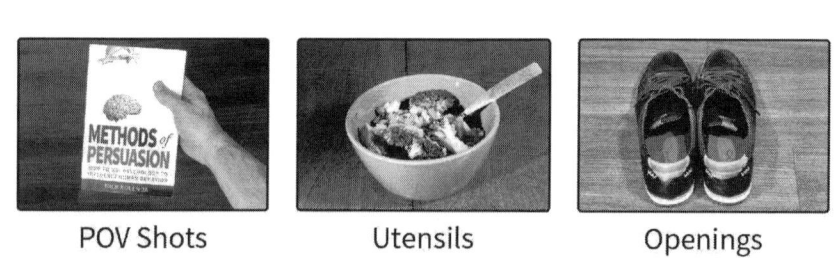

POV Shots   Utensils   Openings

Figure 2D. You can influence action simulations by displaying objects in particular ways.

3. Agents

You simulate your own actions, but you also simulate the actions of other people through *mirror neurons:*

> Every time we are looking at someone performing an action, the same motor circuits that are recruited when we ourselves perform that action are concurrently activated (Gallese & Goldman, 1998, p. 495).

Remember the juggling picture? Viewing that picture activated your own muscles involved with juggling. While determining whether

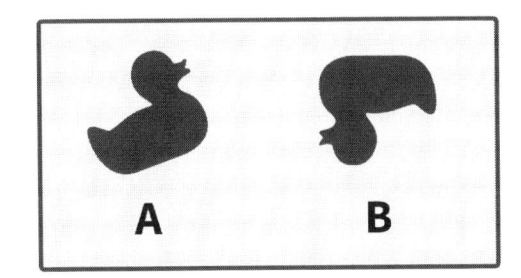

Figure 2E

certain words—catching, kicking, punching—were depicted in the image, you could easily dismiss kicking because your leg muscles weren't activated. However, your arm muscles *were* activated. You needed more time to reflect on punching because you could imagine this motor action more easily (see Zwaan, 1996).

You also immerse yourself *into* agents. But first, let's set the stage with Figure 2E. Determine whether the object on the left matches the two variations on the right.

Done?

Both variations match, but you might have noticed that you needed more time with Variation B. Why was that? It's because you answered that question by mentally rotating the object on the left until it matched the two objects on the right. Variation B took longer because you spent more time rotating.

You follow that same behavior with people. Figure 2F displays my arm in various orientations. For each picture, determine which arm—left or right—is outstretched.

Are you done?

You probably noticed that my left arm is outstretched in all three pictures. More importantly, you determined those answers by immersing yourself into my body. When my back was facing you (Version A), you could answer faster because that orientation

A **B** **C**

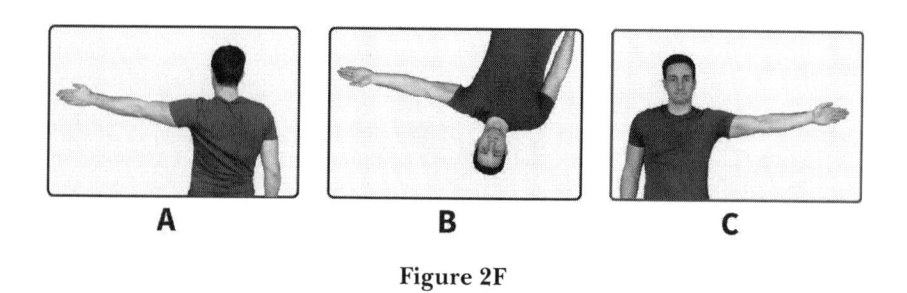

Figure 2F

matched *your* orientation. Your reaction time was slower when I was facing you (Version C) because you needed to orient your body 180° to immerse yourself into my body (Parsons, 1987).

4. Events

Finally, you also simulate experiences. You sometimes hear about visualization: If you visualize an event, you'll perform better.

And that's actually true. Your brain has trouble distinguishing between simulations and real life. In one study, people who imagined eating M&Ms ate fewer M&Ms from a real bowl because their visualization satisfied their desire (Morewedge, Huh, & Vosgerau, 2010). Therefore, visualizing motor tasks (e.g., golf, piano, dance) improves your performance because these simulations replicate actual practice (see Meister et al., 2004).

Now that you understand the four types of mental pictures—*concepts, actions, agents, events*—let's see the implications.

Why Congruence Feels Good

You prefer stimuli that match your simulations. Compare these two deals for pizza:

- 4 small pizzas with unlimited toppings for $24
- 4 small pizzas with 6 toppings for $24

The first deal is economically superior, yet people are more likely to buy the second deal (King & Janiszewski, 2011a).

Notice something about the second deal? What do 4 and 6 have in common with 24? Aha . . . aren't they multiples of 24? Indeed they are. Upon seeing these numbers, you activate 24 on a nonconscious level. This second deal, albeit worse, *seems* favorable because $24 is congruent with your simulation.

But why do you prefer congruent stimuli? You can blame three reasons: *fluency, serendipity,* and *familiarity*.

Fluency

Congruent stimuli (e.g., $24) are easier to process, a concept called *processing fluency* (see Alter & Oppenheimer, 2009).

You can process these stimuli more easily because your simulations create a template for these stimuli (Kok, Failing, & de Lange, 2014). For example, when people saw a photo—face or house—they were quicker to answer questions about these photos (e.g., gender of face, number of stories in house) when they knew which photo was coming. Their anticipation jumpstarted the mechanisms involved with the perception of this object (Esterman & Yantis, 2009).

Serendipity

Remember brightness and sneezing? You couldn't articulate the reason; you just knew that *something* felt right. Language doesn't have a term for this feeling, but we could call it *serendipity*.

Serendipity can explain why *disfluency* can sometimes enhance perception. For example, people preferred a gourmet cheese when the brand name was displayed in a complex font (Pocheptsova, Labroo, & Dhar, 2010). Gourmet cheese, by itself, is unique. By seeing that cheese paired with another unique stimulus—an unusual font—people experienced a sensation that something "felt right."

Familiarity

Our ancestors were more likely to survive if they were cautious with unfamiliar stimuli. Today, congruent stimuli feel good because they feel familiar and safe.

To recap, you create mental pictures of concepts, actions, agents, and events. When you encounter something in the world that matches your internal simulation, you feel a positive emotion because of three reasons: fluency, serendipity, and familiarity. This positive emotion enhances your perception of that stimulus.

But we're still missing something. Why do positive emotions enhance the stimulus? Wouldn't we feel better in general? Why do we attribute these emotions *to* the stimulus?

The final piece of the puzzle is *misattribution*.

Misattribution

Take a guess: Is the letter 'R' more likely to appear in the first or third position of a word?

The correct answer is the third position, but you are more likely to guess the first position (Tversky & Kahneman, 1973). You could brainstorm more words with "R" as the first letter, so you attributed this ease to frequency: *Hmm, I can think of more examples in the first position. It must be more frequent.*

You often feel an ambiguous mixture of emotions, and you attribute these emotions to the wrong source:

- **Truth.** People believed that a statement—*Osorno is in Chile*—was less truthful in a weird font. *Hmm, something about this statement doesn't feel right. It must be false* (Reber & Schwarz, 1999).
- **Value.** Financial stocks perform better when the ticker symbols are easy to pronounce (e.g., KAR vs. RDO). *Hmm, something about this stock feels right. I think the company will do well* (Alter & Oppenheimer, 2006).
- **Liking.** People prefer statements that rhyme (e.g., Life is mostly strife vs. Life is mostly struggle). *Hmm, this statement feels right. I must like it* (McGlone & Tofighbakhsh, 2000).

You prefer congruent stimuli because you feel a positive emotion, and then you rationalize this emotion: *Hmm, this stimulus feels good. Therefore, I must like it because [insert any reason].*

You'll see hundreds of examples in this book, but let's analyze one important caveat.

Caveat: Ignorance of the Source

During sunny weather, people indicate that they are more satisfied with their overall life (Schwarz & Clore, 1983). They feel positive emotions from the weather, and they misattribute these emotions to their life: *Hmm, how do I feel about my life? I feel happy right now. Therefore, I must be happy with my life.*

Interestingly, this effect disappears when asked: How's the weather? That question unveils the true origin of their mood—sunny weather—and they no longer misattribute their feelings to a general satisfaction with life.

So, here's the caveat: Your emotions need to remain ambiguous so that you search for a possible origin behind them.

Remember how you were more likely to buy gourmet cheese in a unique font? You failed to realize that your positive emotions were coming from the design of the packaging, so you concluded that you wanted to buy the cheese: *Hmm, something about this cheese feels good. I must want to buy it.* This effect would disappear if you realized that your emotions were originating from the design: *Hmm, something about this cheese feels good. Oh, it's the graphic design.*

Summary

You "simulate" different scenarios in the world—such as concepts, actions, agents, and events. This imagery doesn't need to be visual; even blind people can run these simulations.

More importantly, you feel positive emotions whenever stimuli are congruent with your mental images, and you attribute these positive emotions *to* the stimuli: *Hmm, this stimulus feels good. I must like it because [insert a reason].*

See Figure 2G for a summary diagram.

Example 1: Shopping Behavior

While buying light roast coffee in a grocery store, you might see a coffee that is congruent with the concept of LIGHT:

It was easier to lift than expected.
It appeared higher on the shelf.
It has minimal text and colors.

Those examples will make sense later. For now, each instance triggers a sensation that something "feels right." At this point, you won't know why something feels right, so your brain will rationalize your emotion by generating hypotheses: *Hmm, this light roast coffee feels good. I must want to buy it because . . .*

. . . *it's less expensive than other coffees.*
. . . *I want to try something new.*

Simulation 23

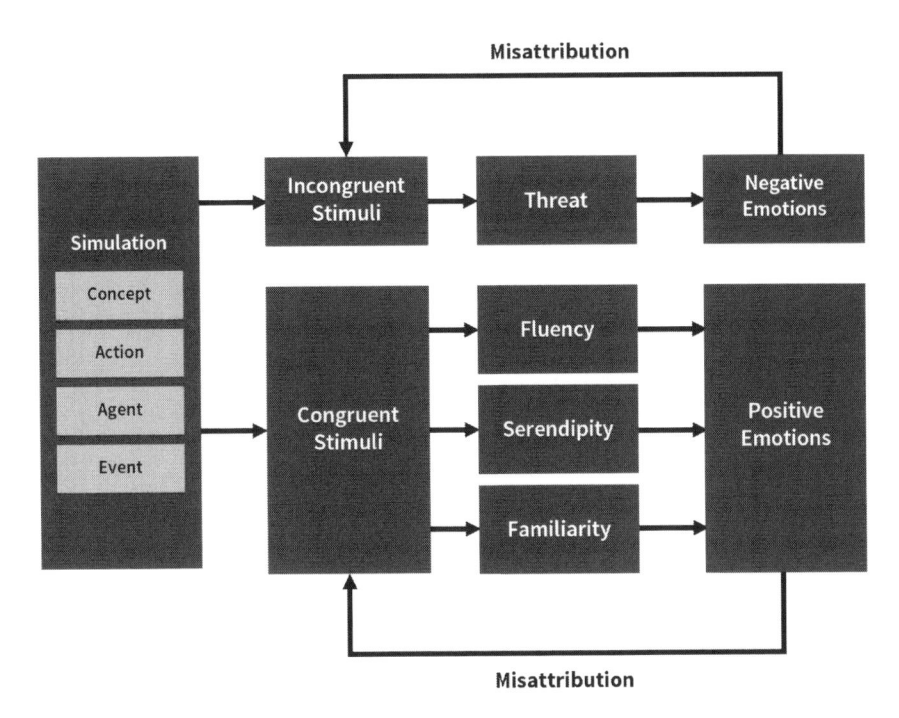

Figure 2G

...Jim enjoys light roast coffee.

You probably generate similar reasons during a purchase. But I have some bad news... those reasons are often meaningless.

Suppose that you found a dark roast coffee that was congruent with your expectations. Your brain might have generated different reasons. *Hmm, this dark roast feels right. I must want to buy it because...*

...I want to splurge on an expensive brand.

...I enjoyed the last dark roast that I bought.

...Pam enjoys dark roast coffee.

Those reasons contradict the other reasons, yet it doesn't matter. Your gut makes the decision, and *then* your brain justifies it.

Example 2: Watching TV

Did you notice something in the previous section? My examples mentioned two people—Jim and Pam—who are main characters on the television show *The Office*. You just experienced one of these outcomes:

- ▶ **Outcome 1.** If you've never seen the show, then nothing happened.
- ▶ **Outcome 2.** If you've seen the show—and if you consciously made the connection to *The Office*—then nothing happened. You'll know why that show became activated in your brain.
- ▶ **Outcome 3.** If you've seen the show—yet you *didn't* notice the connection—then a few things happened. Unbeknownst to you, the show become more activated in your brain. If you were browsing Netflix, you would have felt an ambiguous emotion pulling you toward this show: *Hmm, something about The Office feels right.* You would rationalize this feeling by confabulating reasons to watch this show (e.g., I want to watch something funny . . . the episodes are short . . . I haven't watched it in a while). You encounter those reasons all the time, yet hopefully you see the hidden sources behind them.

You now possess the necessary information to understand all of the implications throughout this book.

In the next part, we'll explore the "primitives" that shaped your perception of the world. The next chapter explores the first primitive—SIZE—and you'll see how everything comes together.

2
Primitives

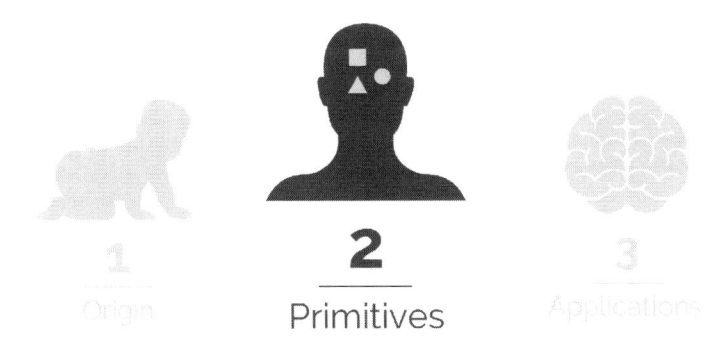

In this part, you'll discover the primitive concepts that shaped your perception of the world.

3

Size

MEMORIZE THESE DIGITS: 6, 9, 7, 8.

Got 'em?

Now, answer quickly . . . did 5 appear in that group? How about 1?

You probably answered "no" for both. And you're correct. However, did you struggle more with 5?

You can blame SIZE.

Abstract Size

You were thrusted into a confusing world of sights, sounds, and bodily sensations. Amidst this confusion, you started noticing that objects could be small or large.

Spatial size became an early concept in your web of knowledge, and you inserted this idea into new concepts, like numbers. When I asked whether 5 or 1 appeared in a group (6, 7, 8, 9), you struggled with 5 because the spatial size of this number resembled the spatial sizes of the other numbers:

> The list is obviously not memorized only as a series of arbitrary symbols, but also as a swarm of quantities close

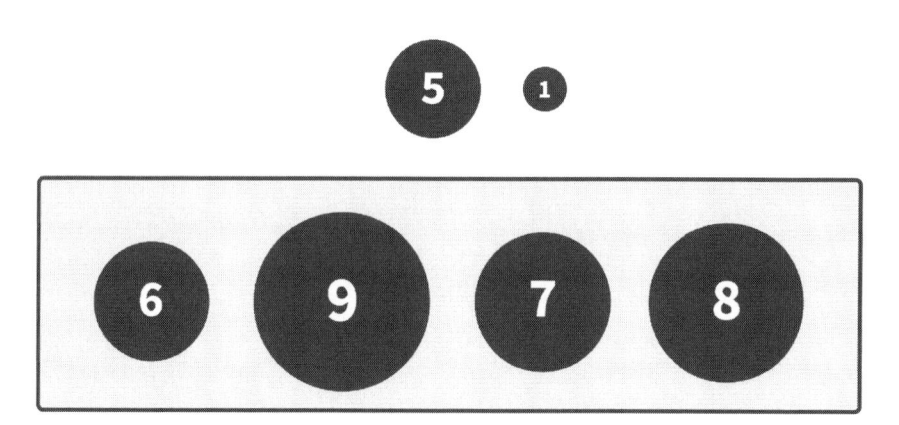

Figure 3A. How your brain conceptualizes numbers.

to 7 or 8—which is why we can immediately tell that 1 is not in the set (Dehaene, 2011, p. 78).

Every time that you conceptualize a number, you construct this mental image with spatial size. Here's another example:

Is 1 less than 5? Is 4 less than 5?

Both numbers are less than 5, but you are slower to determine that 4 is less than 5 because these spatial sizes resemble each other (Dehaene, 2011). Another example:

Is 2 less than 4? Is 26 less than 28?

Both equations share a difference of 2, but you are slower to determine that 26 is less than 28. In terms of spatial size, a change from 26 to 28 is hardly noticeable, but a change from 2 to 4 is *double* in size. You can spot differences in small magnitudes more easily because of the sensory foundation.

We're approaching the precipice of a deep rabbit hole; the next concept plays a huge role in this book. I'll illustrate with a question:

How many miles is the Mississippi River?

Got your guess?

Researchers asked people to draw small or big lines before answering this question, and these lines influenced their guesses. Short lines? The average guess was 72 miles. Big lines? The average guess was 1224 miles (Oppenheimer, LeBoeuf, & Brewer, 2008).

Remember your guess? Was it higher than 72 miles?

Immediately before asking the question, I mentioned a "deep" rabbit hole that plays a "huge" role in this book. Those words activated a general concept of LARGE, which biased your estimate upward.

(In case you're curious, the correct answer is 2,320 miles. Guess we could all benefit from larger cues.)

The takeaway: You inserted a primitive concept of SIZE into many domains, such as lines and numbers. This sensory framework helped you understand all abstract forms of size.

The next section illustrates how this flexible concept of SIZE influences your perception and behavior.

Change the Size of Symbols

Suppose that you are describing a very large size to somebody. How would you say it?

It was huge.

It was h-u-u-u-u-u-u-ge.

The second, right? You naturally enlarge the linguistic sound if you want to communicate a large size.

Many stimuli, such as words, symbolize a particular idea or behavior. Shrinking or enlarging these symbols will influence the perceived "amount" of this idea. Did you notice that all of the chapter titles in this book are very short? You confuse this *shortness* for the amount of work involved: *Hmm, how long will it take to read this chapter. Something feels short. It should be easy to get through.*

You find a similar effect with numbers. For many years, calories were displayed in a small font on the back of food packages, as if they were negligible and unimportant. But recently, the FDA started requiring manufacturers to display a larger digit so that people are less likely to overconsume: *Hmm, how many calories are in this food. Something feels big. This must be a lot of calories.*

Or consider prices. Prices seem cheaper when they are visually smaller: *Hmm, how much does this cost? Something feels small. The price must be small* (Coulter & Coulter, 2005).

Whenever you encounter these symbols—words, calories, prices—you paint a mental image of the intended concept. Therefore, the sensory traits of these symbols can distort the resulting images. A visually larger number will create a mental picture in which the abstract size is larger.

And it's not just visual size. Consider these prices:

- $1250
- $1,250

Both prices are the same, right? Technically, yes. However, the second price (with the comma) seems larger. Do you see the culprit?

Sure, the comma enlarges the surface area of the price—but it also enlarges the linguistic sound:

- **$1250:** Twelve fifty (3 syllables)
- **$1,250:** One thousand two hundred fifty (8 syllables)

Customers confuse linguistic size for numerical size because both forms of size are constructed from the same primitive ingredient of SIZE (Coulter, Choi, & Monroe, 2012).

Change the Units

The number 1 seems much lower than 85,000, right? But that's not necessarily true . . . 85,000 seconds is smaller than 1 day.

Units of measurement determine the meaning of numbers, yet research shows that we consistently overlook these conversions (Shen & Urminsky, 2013). Heck, even now, you *still* probably feel like 85,000 seconds is larger. You experience that intuitive feeling because your brain is translating these numbers into a spatial size.

Therefore, you should choose units strategically. In any scenario, you can describe information with different units: 80% of people . . . 8 out of 10 people . . . 800 out of 1000 people. Why not choose units that communicate the ideal size? Cancer seems riskier when it kills 1,286 out of 10,000 people, compared to 24.14 out of 100 people (even though the death rate is lower; Yamagishi, 1997).

Or consider your spending while traveling abroad. You are influenced by the nominal values of currencies: You underspend when the foreign currency is larger (75 Indian Rupee = 1 US Dollar), but you overspend when the domestic currency is larger (0.4 Bahraini Dinar = 1 US Dollar; Raghubir & Srivastava, 2002).

Be careful with marketers who mix different units. Suppose that you're choosing a monthly subscription for food delivery:

- 5 meals each week ($65/week)
- 7 meals each week ($85/week)

The second plan has 2 more meals for an extra $20. But oh, marketers are sneaky devils. They can tweak these units:

- 20 meals each month ($65/week)
- 28 meals each month ($85/week)

Whaddya know . . . now it *seems* like you're getting 8 meals for a $20 difference (Burson, Larrick, & Lynch Jr, 2009).

Change the Reciprocal

If you've read 5% of this book, then you have 95% remaining. Which variation would motivate you to continue reading?

Researchers investigated this question by distributing loyalty cards at a sushi restaurant. They compared two versions: accumulating progress (e.g., collect 10 stamps) or remaining progress (e.g., get hole punches in 10 existing stamps). Turns out, customers preferred whichever card possessed a smaller magnitude (see Figure 3B). Collecting stamps was more effective toward the beginning with only a few stamps, but hole punches were more effective toward the end with only a few stamps remaining (Koo & Fishbach, 2012).

Two reasons. First, customers confused the smaller magnitude with less work: *Hmm, how much work is involved? Something feels small. It must be less work.*

Collect 10 Stamps — Accumulated

Get 10 Hole Punches — Remaining

Figure 3B

Second, remember the magnitude effect? When you move from 27 to 28, the change is small. When you move from 1 to 2, however, the change is huge . . . it's double. With two stamps, a new stamp is 50% of your progress. That's a big jump. On the other hand, with eight stamps remaining, a stamp removal—the same implication—feels less impactful.

Right now, you are motivated to continue reading with 5% progress in this book. Every new page feels more impactful. Once you cross the halfway point, however, you will prefer remaining progress because this percentage will now be a smaller magnitude.

Prime a Magnitude

What happens if you can't change the symbol, unit, or reciprocal? You can use priming. A large magnitude in one context (e.g., big lines) can enlarge the magnitude of a subsequent context (e.g., Mississippi River).

Sometimes, however, priming causes the reverse effect: A large magnitude in one context can shrink the perceived magnitude of a subsequent context. In a field study, a street vendor sold music CDs, while an adjacent vendor sold sweatshirts (and alternated the price between $10 or $80). The researchers found a *contrast* effect: When the sweatshirts were expensive, the CDs seemed cheaper (Nunes & Boatwright, 2004).

So, what's going on? Does priming cause *contrast* or *assimilation*? Suppose that you buy a $10 product, and you see these words: Join 2,387 happy customers. Will the price seem lower or higher?

The answer can be found in another primitive—OBJECT—which we'll discuss in the next chapter.

Summary

Spatial size appears in every corner of the world in different modalities (e.g., sight, touch). You learned this concept early in life, and you inserted this idea into many abstract domains. Ultimately, all forms of size are built with this primitive of SIZE.

Want to influence the abstract size of something? You can simply tweak the symbolic depiction. With numbers, you could:

- Adjust the visual size of the digit
- Choose an appropriate unit
- Emphasize a reciprocal variation

Other examples:

- **Product Images.** Do you sell products online? How big should your images be? It depends on the traits that your customers are seeking. Do they want a laundry basket that holds a lot of clothes? Enlarge your images. Do they want a laundry basket that is lightweight, compact, or storable? Reduce the size of your images.
- **Product Features.** Selling a camera? You could describe the image resolution by mentioning the number of pixels along the diagonal (e.g., 2,000 pixels). But why? You should describe the image resolution with total pixels (e.g., 4,000,000 pixels) because the larger numeral is more persuasive (Hsee, Yang, Gu, & Chen, 2008).
- **Product Warranty.** An 84-month warranty is more persuasive than a 7-year warranty (Pandelaere, Briers, & Lembregts, 2011).
- **Daily Price.** Marketers often convert a price into the daily

equivalence, such as $2.50/day. This *smallness* distorts the mental image of the total price.

▶ **Exercise.** Suppose that you're doing 50 pushups. In which direction should you count: upward (1 to 50) or downward (50 to 1)? I'd argue for downward. If you count upward, your body will be physically tired when you reach larger numbers. Those magnitudes (e.g., 47 . . . 48 . . . 49) will feel heavier. If you count downward, the final numbers (e.g., 4 . . . 3 . . . 2) will feel smaller and lighter, giving you the extra push. Later, you'll learn about WEIGHT (and why larger numbers *are* heavier).

4

Object

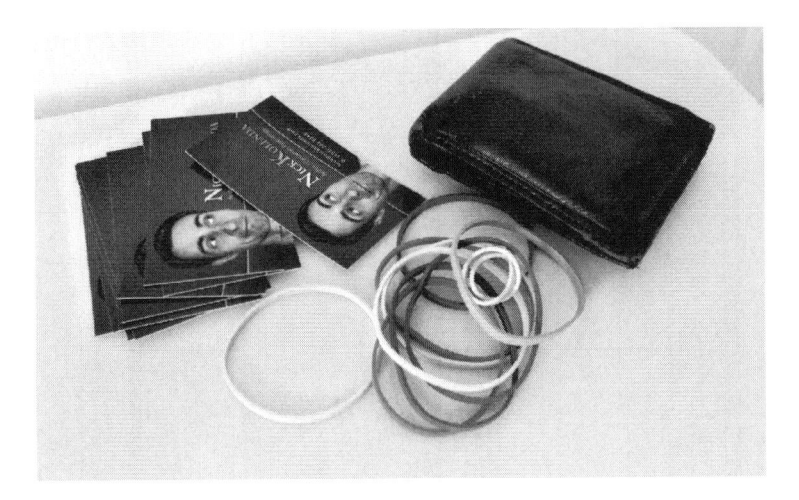

Figure 4A

ON MY DESK are three items: my wallet, business cards, and rubber bands. In this chapter, I'll argue that I'm more likely to buy something after seeing this cluster. Any guesses why? The answer involves the OBJECT primitive.

The sensory world is filled with objects. In order to navigate this cluttered mess, you need to group everything.

Enter: *Gestalt principles.*

Gestalt Principles

You are holding this book. So, how many objects are you holding?

Did you guess a single object? But why? This book contains hundreds of pages. Why didn't you count these pages?

Or maybe you're reading the electronic version. You might have conceptualized your device to be a single object—but again, why? Doesn't your device contain smaller parts and components?

In reality, the term "object" is entirely flexible. Even though you perceive boundaries around certain objects—books, devices, etc.—these boundaries are malleable. Instead, you are surrounded by a *continuum* of stimuli that you can categorize in limitless ways. For example, read the following word:

OBJECT

Seems like a cohesive entity, right? But it's not—we could transform this entity into multiple objects by separating the letters:

O B J E C T

Voila . . . it's now six objects. You can perceive *anything* to be an object; it just depends on your perception.

So, when do you perceive something as a cohesive unit? Researchers uncovered a collection of guidelines called Gestalt principles (see Figure 4B for examples). This book seems like a single object because all of the pages are connected in close proximity.

But again, *anything* can be an object. Remember the items on my desk? Weren't they connected in close proximity, too? Indeed, your brain was perceiving this cluster of items as a single object, which has profound implications, as you will soon discover.

Proximity   Similarity   Containment

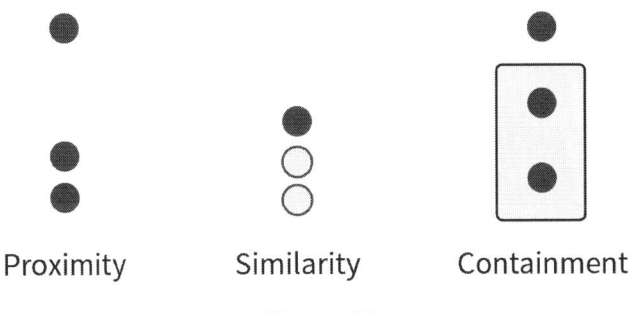

Figure 4B

Indexing Your Attention

Here are two shapes: red circle and green square. *Remember them.* We'll revisit them later, and you'll see why.

You are looking at this book, but you aren't paying attention to all of these sentences inside your field of vision. You are focused in this sentence. And now this sentence.

I'll refer to this mechanism as *indexing*: You can view or ignore certain objects inside the same field of vision.

Think of your attention as a human hand, which has fingers that you can attach to objects. It's a theory called FINST (Pylyshyn, 1989).

Let's walk through an example.

Find three objects nearby. I'm looking at a book, highlighter, and microphone. Once you find three objects, individuate them. Fixate on each object so that you distribute your attention equally among them. Don't focus on the group. And don't fixate on items sequentially. Divide your attention into equal streams that latch onto each object. It sounds weird in writing, but you'll understand when you try it.

Once you've attached those fingers, notice your level of effort. How does it feel? It should feel normal and doable.

Now, find a fourth object nearby. Got it? Distribute your attention equally among all four items. How does it feel now? Tougher, right? But hopefully still manageable. Research shows that we have a capacity of 4 indexes or fingers (Scholl, 2001).

You have now reached your capacity, so let's test that limit. Find a fifth object nearby, and add a new stream. At this point, the difference should be clear. When you distribute your attention equally among all five objects, your attention becomes lopsided. You'll start focusing more heavily on some objects, while neglecting others. In order to fixate on a new object, you need to detach one of the active fingers:

> When all available indexes are already assigned to objects, a new object can be indexed only by first de-assigning one of the active indexes, flushing its bound features, and reassigning it to the new object (Scholl & Leslie, 1999, pp. 28–29).

Even though you're limited to 4 indexes, you can overcome that limitation by grouping objects—an effect that is similar to the grouping of phone numbers (781-314-3425 vs. 7813143425).

The takeaway: You group stimuli in the external world by placing them into various slots of your attention. Everything inside one of these slots will seem like a cohesive object.

The last piece of the puzzle is feature integration.

Feature Integration

You see a stranger walking toward you. Suddenly he starts talking to you, and you realize something: It's not a stranger—it's your friend, Gary. In this situation, you wouldn't look around, wondering where the stranger went. Or how your friend magically appeared.

You would recognize that your friend was, indeed, the stranger the whole time.

You perceived the unity of Gary because you created an *object file* upon seeing him (Kahneman, Treisman, & Gibbs, 1992). You created a symbol that inherited the spatiotemporal features of Gary, rather than the visual or semantic features. In other words, you individuated Gary before you fully recognized him.

At this point, you start activating the individual features of an object before merging them into a cohesive perception. For example, when you see an apple, you encode "redness" and "roundness" before combining these features into the coherent perception of an apple (Hazeltine, Prinzmetal, & Elliott, 1997).

Earlier, I mentioned two shapes with different colors. Can you name them? Although you might have guessed correctly—it was a red circle and green square—those four traits (e.g., square, round, red, green) were floating aimlessly in your brain, detached from their true counterparts. Many people mistakenly guess a red square and green circle (Treisman & Schmidt, 1982).

To recap, you can perceive *anything* to be an object, including a group of random items on a desk. When you distinguish something as an object—which, again, could be multiple objects—you assign a single object file. Then, you activate the features of *everything* inside this group, and you insert these features into the object file.

Still with me?

Don't worry if you feel lost. The next section will explain the main implication, which is easier to grasp.

Convergent Processing

Suppose that I hid a prize in a paper bag. Which assortment would you want to choose from in Figure 4C?

Figure 4C

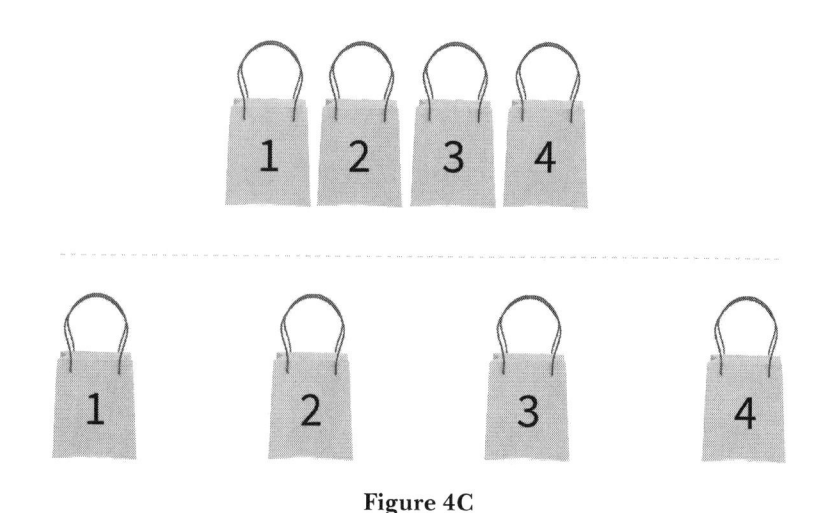

Most people say the top assortment in which the bags are closer. Doesn't it feel more enticing, as if you'd be more likely to win?

So, what's going on? You are encountering the same effect from earlier. Recall this word:

OBJECT

Again, this word seems like a cohesive object—that is, until you disperse the letters away from each other:

O   B   J   E   C   T

In Figure 4C, you are grouping the top assortment of bags as a single object because of their proximity—and thus, you generate a single object file. Then, you activate the features of *all* bags, and you insert these features into the single object file. Do you see the issue? One bag has a prize. Therefore, you insert this feature into the single object file. Other bags will inherit this feature because they are part of the same object file. If one bag has a prize, *all* bags will have the prize because they belong to the same entity.

Moving forward, I'll drop the technical details. I'll simply refer to this mechanism as *convergent processing*: Whenever you group items as a unit, you merge the features. According to your brain, this group of items *is* a cohesive object.

So, what's special about the items on my desk? How could they influence my spending behavior? You can blame the rubber bands. When people are exposed to words about flexibility—*rubber, elastic, flexible*—they become open-minded because these words activate an abstract flexibility (Hassin, 2008). On my desk, the rubber bands (and flexibility) can merge with my wallet. If I look at my wallet while contemplating a purchase, my spending behavior will seem more flexible during this moment.

And sure, that example might be a bit colorful—but this book will expand on the empirical support in future chapters.

For now, here are some applications with prices and people.

Prices

In the previous chapter, I asked whether $10 would seem lower or higher near these words: Join 2,387 happy customers.

Can you solve it now?

It depends on the grouping. If you group $10 and 2,387 inside the same FINST, perhaps because these numbers are close together, then the $10 price will seem higher. On the other hand, if you perceive these numbers to be separate, you experience a reciprocal effect—divergent processing—in which you polarize the magnitudes away from each other (so the $10 price will seem lower).

Nearby words can also trigger this effect: An inline skate seemed cheaper when the price appeared next to the words "low friction." Conversely, the price seemed higher next to the words "high performance" (Coulter & Coulter, 2005).

People

Convergence occurs with people, too. Suppose that you're standing next to somebody who is very attractive. Will you look better (convergence) or worse (divergence)?

Again, it depends on the grouping. Gestalt principles will increase your attractiveness: You're standing near your friend . . . your shirt is the same color . . . you're sitting on the same couch.

Even a conceptual relationship works. Consider a neutral person who is standing near an attractive person. You evaluate this person's attractiveness in accordance with the relationship to the adjacent person. Friends? The neutral person seems *more* attractive because of convergence. Strangers? The neutral person seems *less* attractive because of divergence (Melamed & Moss, 1975).

Later in the book, you'll see how convergence shapes fundamental human behaviors, such as empathy.

False Equality

Finally, objects also display a false sense of equality. For example, researchers offered candy from three bowls:

- **Bowl 1.** Candy A
- **Bowl 2.** Candy B
- **Bowl 3.** Candy C and Candy D

Since Bowl 3 contained two types of candy, people should have chosen from that bowl more often to account for the difference. But they didn't. They conceptualized only three objects—Bowl 1, Bowl 2, Bowl 3—and they chose from each bowl roughly 33% of the time (Fox, Ratner, & Lieb, 2005).

You can apply this insight to influence choices. A school cafeteria might offer two categories of food:

- Healthy
- Unhealthy

These categories are detrimental because the unhealthy food shares equal footing with healthy food. How could we nudge students to choose healthier food? We could expand the healthy group into more objects:

- Healthy—Vegetables
- Healthy—Fruits
- Unhealthy

Now, the healthy foods comprise a larger portion of the group. Students will choose from this category at a higher rate.

You also generate your own categories of objects. Imagine that you see three options of yogurt: blueberry, strawberry, and strawberry cheesecake. If you are buying 12 yogurts, your choices will depend on the way you group these options.

If you categorize on *flavor*, you have three options: blueberry, strawberry, and strawberry cheesecake. You might buy four of each.

However, suppose that you categorize on *fruit*. You might combine strawberry and strawberry cheesecake into a single "Strawberry" category. Now, you have two options: strawberry and blueberry. You might buy six of each.

Or consider *healthiness*. You might combine strawberry and blueberry into a "Healthy" category. Now you have two options: healthy and unhealthy. Again, you might buy six of each.

Summary

The term "object" is inaccurate. You aren't surrounded by discrete objects; you are surrounded by an endless continuum of stimuli, a continuum that you can group in limitless ways.

Gestalt principles are the default mode of grouping. Items will seem like a single object if they are similar, close together, or enclosed inside the same boundaries.

Other examples:

- **Assortment of Bags.** Remember the bags with a prize? You chose from the tight assortment because you merged the prize with all bags. The reverse happens with negative traits: People preferred a spacious assortment of ketchup bottles if one bottle had a defective lid (Mishra, 2008).
- **Perception of Meals.** Suppose that your friend orders a cheeseburger, and you order a salad. You will polarize these calories: The salad will seem like fewer calories, while the cheeseburger will seem like more calories. However, if you order a salad and cheeseburger as a single meal, you merge the calories: The cheeseburger seems healthier because it inherits healthiness from the salad (Wilcox, Vallen, Block, & Fitzsimons, 2009).
- **Spending.** Suppose that you're standing in line to buy a $10 meal for lunch. You look inside your wallet, and you realize that you lost a $10 bill. Would you still buy lunch? Probably, right? But now, suppose that you bought the $10 meal—and upon sitting down—you dropped everything to the floor. *Ugh.* The meal is no good. In this situation, would you buy another lunch? Most people wouldn't buy another lunch even though they lost the same amount of money. The second scenario feels more painful because you categorized that money within a "lunch" budget (see Thaler, 1985).

5

Location

DURING THE 2004 presidential debate, George Bush gestured with his right hand to convey positive statements (e.g., "and they'll continue to keep their checks"). During the 2008 debate, Barack Obama gestured with his left hand for positive statements (e.g., "you can keep your health insurance").

An arbitrary coincidence? Perhaps . . . or perhaps not. I'll give you a few paragraphs to brainstorm an explanation.

In the previous chapter, you learned that babies start discovering objects in the sensory world. Well, these objects appear in different locations—UP, DOWN, LEFT, and RIGHT. And you insert these locations into abstract ideas.

You already learned a few connections, such as UP and POWER. Here are some additional examples to jog your memory:

- ▶ Brands seem more powerful when their logos appear toward the top of packaging (Sundar & Noseworthy, 2014).
- ▶ People with dominant personalities are quicker to detect changes in their upper field of vision (Moeller, Robinson, & Zabelina, 2008).
- ▶ Elongated products, such as mascara, seem more powerful when displayed vertically (Van Rompay, De Vries, Bontekoe, & Tanja-Dijkstra, 2012).

Perhaps you could experience this effect right now. Which country is bigger: Congo or Zambia? And how confident are you?

While answering that question, people felt more confident on the 8th floor of a building (compared to the 2nd floor) because they felt more powerful (Sun, Wang, & Li, 2011). Where are you located right now? Your spatial location might have distorted your level of confidence in the answer.

Size

You also learned that UP is MORE, in which large numbers move in an upward direction. Turns out, you conceptualize numbers along a horizontal spectrum, too.

Estimate the midpoint of this line:

xxxxxxxxxxxxxxxxxxxxxxxxxxxx

Your guess was probably accurate. However, your answer would have changed with these lines:

2222222222222222222222222222

8888888888888888888888888888

In your brain, numbers are mapped along a horizontal spectrum. Small digits (e.g., 2) bias your estimate leftward, whereas large digits (e.g., 8), bias your estimate rightward (Fischer, 2001).

Interestingly, some people (e.g., Iranians) show the reverse pattern: Numbers get larger toward the left. Can you spot the reason? Or how we derived this connection in the first place?

The answer: Reading directionality.

You believe that SIZE moves from left to right if you read in a language that moves from left to right. Iranians establish the reverse direction because they read from right to left. For simplicity, I'll assume a rightward directionality in this book.

And finally, numbers aren't the only culprit. Recall that SIZE is a flexible concept that you insert into many domains. Consider emotional magnitude. If you conceptualize large magnitudes toward the right, then you should be quicker to identify a big smile, a large magnitude of happiness, toward the right. And, sure enough, that's what happens (Holmes & Lourenco, 2011).

Stimuli also appear smaller or larger in these locations. Calories, for example, seem smaller toward the left of a food package (Romero & Biswas, 2014). Part of this effect originates from a visual fulcrum in the design:

> . . . because our eyes enter a visual field from the left, the left naturally becomes the anchor point or "visual fulcrum." Thus, the further an object is placed away from the left side (or the fulcrum), the heavier the perceived weight (Deng & Kahn, 2009, p. 9).

You can see this illusion in Figure 5A; a price seems heavier toward the right of a price tag.

The vertical spectrum is more convoluted. Although SIZE gets

Figure 5A

3 8 S

Figure 5B

larger as it moves upward, heavy objects also pull downward. So, you also associate LARGE and DOWN.

Look at the font style in Figure 5B. Everything looks symmetrical, right? Well, don't be fooled. When you flip the graphic upside down, you'll discover that the bottom halves were actually larger. Designers sometimes add more visual weight in bottom areas because your brain expects more weight in these locations (Arnheim, 1974).

Cardinal Directions

Do you ever confuse east and west? I feel like a child whenever I use the mnemonic "Never Eat Soggy Waffles" (which is every time).

Despite that difficulty, do you ever confuse north and south? Probably not, right? Why is that?

Turns out, you scaffolded cardinal directions of north, south, east, and west onto UP, DOWN, LEFT, and RIGHT. You never confuse UP and DOWN, but you sometimes confuse LEFT and RIGHT, so east and west inherited this difficulty.

But why do you confuse LEFT and RIGHT? Horizontal directions are troublesome because they feel identical: Turning in one direction will eventually bring you to the opposing direction. Vertical directions are easier to distinguish: You can easily go DOWN, but it takes effort to move UP. Turning in this spectrum is more difficult—see Figure 5C for the results of my scientific study.

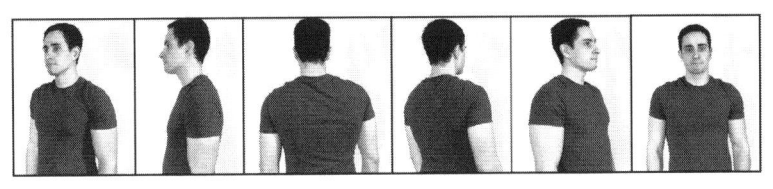

Horizontal Rotation

Vertical Rotation

Figure 5C

Good

When you feel good and energized, what do you do? You stand up. You walk around. When you feel sick or sad, what do you do? You slump. You lie down.

Therefore, you start associating UP and GOOD. Consider this passage from a study called *Why the Sunny Side is Always Up*:

> Objects that are up or high are often considered to be good, whereas objects that are down or low are often considered to be bad. In the Bible, for example, the righteous go "up" to Heaven, whereas sinners go "down" to Hell. In the media, movie critics give good movies "thumbs up" and bad movies "thumbs down" ... People who smoke marijuana "get high," but when the euphoria diminishes, they "come down," and happy people feel "up" whereas sad people feel "down" (Meier & Robinson, 2004, p. 243).

Indeed, UP and GOOD are connected in your brain: Activating one concept will activate the other.

- ▶ Positive pictures seemed higher on a screen (Crawford, Margolies, Drake, & Murphy, 2006).
- ▶ Positive memories were easier to remember while moving marbles to a higher location (Casasanto & Dijkstra, 2010).
- ▶ Depressed people were quicker to detect changes in their lower visual field (Meier & Robinson, 2006).

And it gets more interesting. You insert UP into many abstract domains, so these concepts—such as NORTH—will inherit this valence. When researchers asked people to mark their ideal place to live in a fictional city, they consistently chose northern locations (Meier, Moller, Chen, & Riemer-Peltz, 2011).

You find similar effects in the horizontal spectrum. Meet Barbara in Figure 5D. She hates zebras, but she absolutely loves panda bears. In which box would you place each animal?

Most people place "good" animals on the same side of their dominant hand: Right-handed people placed good animals on the right, while left-handed people placed good animals on the left (Casasanto, 2009).

In another study, people memorized the locations of 32 fictional events on a map. Some events were positive, negative, and neutral:

- ▶ **Positive.** Six kittens are rescued from a tree.
- ▶ **Negative.** Child attacked by grizzly bear.
- ▶ **Neutral.** Middle school track meet rescheduled.

When asked to replicate those locations on a blank map, people showed an upward bias for positive events (and a downward bias for negative events). They also showed a horizontal bias: Right-handed

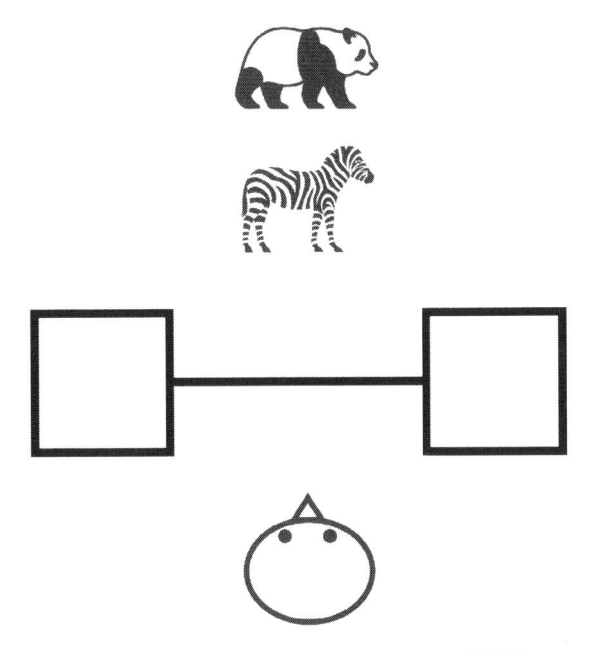

Figure 5D. Adapted from Casasanto (2009)

people believed that positive events happened farther to the right, and they believed that negative events happened farther to the left. Their bias even followed the degree of handedness: People who were *more* reliant on their right hand placed the positive events *farther* to the right (Brunyé, Gardony, Mahoney, & Taylor, 2012).

This insight can shed light on my question at the beginning of the chapter. Why did George Bush convey positive statements with his right hand? Because he is right-handed. Barack Obama conveyed positive statements with his left hand because he's left-handed (Casasanto & Jasmin, 2010). The same effect occurred with John Kerry (right-handed) and John McCain (left-handed).

In fact, what if handedness could influence your political beliefs? Perhaps Bush has right-wing beliefs because he's right-handed, while

Obama has left-wing beliefs because he's left-handed. It sounds far-fetched, but it's plausible. When left-handed people discover "left-wing" beliefs, they might perceive a greater familiarity in that spatial label—and they might confabulate reasons to justify those beliefs. Same with right-handed people and right-wing beliefs.

That bias, if any, would likely be small, but you might find more left-handed liberals (and right-handed conservatives).

Canonical Locations

Recall canonical prototypes. Most people draw a teacup with a similar style because they possess a prototypical version of a teacup.

Canonical prototypes contain specific traits, such as locations. For example, you can read the word "attic" more easily when it appears above the word "basement" because these locations match the canonical locations (Zwaan & Yaxley, 2003).

Heck, you can feel this effect right now by reading the words "left" and "right." You would have struggled to read the previous sentence if I had reversed those spatial locations—right and left—like I did now.

The key idea: You prefer stimuli in the locations that you expect to find them.

Remember how size gets larger toward the right? Time moves in a similar direction: The past is on the left, while the future is on the right. In Chapter 1, you saw a picture of myself as a baby and adult. My baby picture was on the left, while my adult picture was on the right. Imagine this graphic with those positions reversed.

Feels weird, doesn't it?

You prefer depictions of time in canonical locations. Weight loss ads are more effective when the heavier model—the "past" image—appears on the left (Chae & Hoegg, 2013).

Importance

You might remember a TV show called *The Weakest Link*. Contestants answered questions as a group, and they voted off people who dragged them down. Researchers noticed that contestants in the middle were more likely to win (Raghubir & Valenzuela, 2006). Positions were randomized, so nothing was unique about the center. And they didn't answer more questions correctly. So, what was happening?

Upon seeing an object, you place this object in the center of your vision. It's called the *central fixation bias*. Across your life, you gradually associate CENTER and IMPORTANCE because important things are literally front and center. People overlooked the errors of middle contestants because those contestants seemed more important.

This tendency can also influence your everyday choices. Suppose that you're choosing an option from Figure 5E. Your eyes will immediately gravitate toward the center. Once you evaluate those central options, you still need to view the options on both sides. So, you view the leftward options. Then you view the rightward options. And perhaps you look side-to-side a few more times before resting your gaze in the center for balance.

That process seems innocent, but it biases your choice toward the central options. How so? When you tally the number of eye fixations, you'll notice that you viewed the central options more often. Every time that you shifted your gaze to the outermost edges, you crossed over the middle options. Those multiple fixations trigger a vicious cycle: The more you look at an option, the more you like it . . . the more you like an option, the more you look at it (Atalay, Bodur, & Rasolofoarison, 2012).

If you want somebody to choose a specific option, you should place this option in the center.

Figure 5E

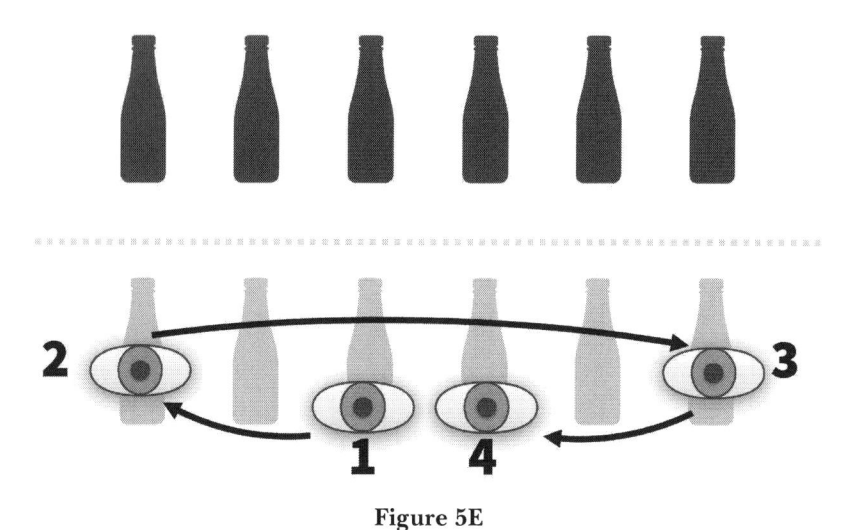

Summary

You insert the sensory trait of LOCATION into many ideas—such as SIZE, GOOD, and IMPORTANCE.

- ▶ Size gets bigger toward the right and upward
- ▶ Good stimuli appear on your dominant side
- ▶ Important things are front and center

You can place stimuli in these locations to instill those traits. For example, if you are reading the paperback version of this book, you are staring at two sides: left and right. Perhaps right-handed people view the right-side pages to be more important.

Other examples:

- ▶ **Eye Gaze.** While thinking of numbers, you look in the respective directions (Fischer Castel, Dodd, & Pratt, 2003). Researchers asked people to name 40 numbers from 1 to 30 as randomly as possible, and they used the *direction* and *degree* of eye change to successfully guess those numbers (Loetscher, Bockisch, Nicholls, & Brugger, 2010).

- **Canonical Location.** Participants preferred a lamp that matched the canonical location of time (i.e., antique lamp on left; modern lamp on right; Chae & Hoegg, 2013).
- **Hiring.** Given two columns of stimuli, such as job applicants, you ascribe more value to the items on your dominant side (Casasanto, 2009).
- **Slogans.** Emotional slogans are preferred toward the bottom of advertisements (Cian, Krishna, & Schwarz, 2015). Perhaps we can blame a hidden magnitude: The emotional weight sinks downward, so this bottom location "feels right." This idea will make sense when we discuss "heavy" emotions later.

6
Distance

THINK OF SOMETHING that you want to do in five years.
Now, some unrelated questions. For each item below, choose the description that feels right.

Making a List
a. Getting organized
b. Writing tasks down

Paying Rent
a. Maintaining a place to live
b. Writing a check

Eating
a. Chewing and swallowing
b. Getting nutrition

We'll revisit these answers later.

For now, this chapter merges all of the past primitives: When two OBJECTS reside in different LOCATIONS, there will be a SIZE of space between them.

DISTANCE is the SIZE between objects.

Consequently, all of the effects with SIZE will be inherited by DISTANCE, as you'll see next.

Abstract Distance

Remember how SIZE is a flexible concept? All types of sizes—visual size, linguistic size, numerical size—are built with the same primitive ingredient. Same with DISTANCE.

Consider this simple notation of time: 6–8pm.

That brief notation has multiple forms of distance: time, numbers, spatial distance between those numbers. All of these distances are based on the same ingredient.

You also insert DISTANCE into social relationships. You can spot these metaphors in language:

> You might be CLOSE to your friends and family, even though you act DISTANT occasionally.

Indeed, people were quicker to read the word "friend" while physically near this word. While farther away, they were quicker to read "enemy" (Bar-Anan, Liberman, Trope, & Algom, 2007).

Again, all of these distances—temporal distance, numerical

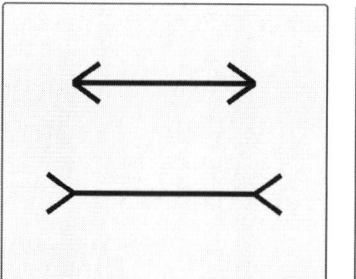

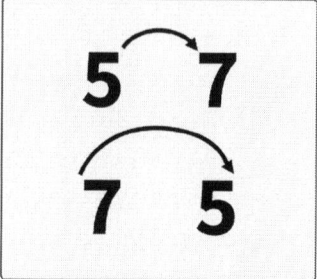

Figure 6A

Distance 61

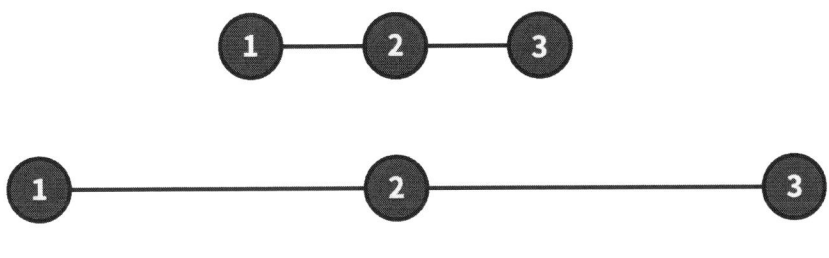

Figure 6B

distance, visual distance, social distance—are constructed with the same primitive ingredient of DISTANCE.

Therefore, you confuse these distances. Think of a store that is far away from you. Thinking of one? Right now, you believe that you have less in common with the sales clerk. Why? You are confusing spatial distance with social distance (Zhang & Wang, 2009).

Or suppose that you need advice. You will prefer the advice of a stranger if the decision involves the distant future, yet you will prefer the advice of a loved one if the decision involves the near future. Again, you are confusing social distance with temporal distance (Zhao & Xie, 2011).

Just like you did with SIZE, you can influence the perception of DISTANCE by tweaking the symbolic depictions. Recall this notation of time: 6–8pm. You could make this time period seem longer by visually expanding this distance, say 6 — 8pm.

It even happens with the Mueller-Lyer illusion. In Figure 6A, the bottom line *seems* longer because your eyes move across a longer distance. Suppose that a kitchen utensil is $7, but is on sale for $5. Marketers could make the discount seem larger by positioning the $7 on the left, which will trigger the same illusion (see Coulter, 2007).

Finally, suppose that you're completing an online form. You typically see symbols for each step (Figure 6B). In the bottom version, each step is spatially farther away, so you believe that each step takes longer to complete.

Ultimately, you possess a primitive concept of DISTANCE that you insert into many domains. Anything can feel subjectively near or far from you.

Simulations

Find an object that is far away from you. I'm gazing out the window of my 11th floor apartment, and I can see a distant drawbridge. I can see the overall shape, but I can't distinguish the concrete details. Are you experiencing the same haziness?

Now, find an object within your reach. Study it closely. I picked up my wallet, and I could feel the texture. I could see the stitching. I could see discoloration. None of these details would have been visible from far away.

You experience this discrepancy in the sensory world: You can see the details of nearby objects, but you can only see the gist of distant objects. And you insert this discrepancy into DISTANCE.

For example, imagine reading this book tomorrow. Now, imagine reading this book in one year.

Each simulation—tomorrow vs. next year—should have generated the same vividness, right? After all, it's only a mental picture. However, that's not the case. Your simulation of tomorrow was more vivid:

> . . . representations of temporally close events (both past and future) contained more sensorial details, were associated with a clearer representation of contextual information (location, time of day), and generated a stronger feeling of re-experiencing (or pre-experiencing) than representations of temporally distant events (D'Argembeau & Van der Linden, 2004, pp. 17–18).

You insert DISTANCE into time, so you insert all of the pertinent traits of this idea: Objects appear hazy across a far spatial distance, so they appear hazy across a far temporal distance.

In fact, your imagery contains less color, too. In one study, people imagined a housewarming party in the future. Researchers gave them a blank drawing of the party with 10 colored pencils: five pencils were chromatic colors, and the other five pencils were variations of gray. When they were told the party was happening in the distant future, they used more variations of gray because their mental imagery was less detailed and colorful (Lee et al., 2016).

To recap, spatial distance has a constant discrepancy: You see the details of nearby objects, but you see the gist of distant objects. You insert this discrepancy into all abstract notions of distance.

Over time, these perspectives can also influence the way you think about objects; it's a principle called *construal level*.

Construal Level

What do you see in Figure 6C? From far away, you will see the gist (the H). Up close, you will see the details (the S's).

- **High Construal.** You see the gist of "H"
- **Low Construal.** You see the details of "S"

From far away, you construe information in broad terms. Consider a hypothetical event, such as locking a door. If this event is happening next year, you construe this behavior in broad terms (e.g., securing a house). If this event is happening right now, you construe the specific terms (e.g., putting a key in a door; Liberman & Trope, 1998).

Sound familiar?

THE TANGLED MIND

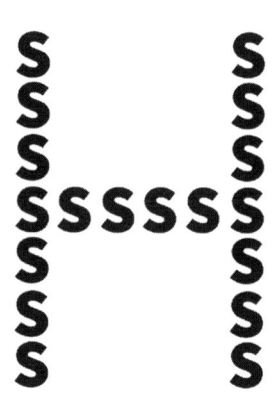

```
S        S
S        S
S        S
SSSSSS
S        S
S        S
S        S
```

Figure 6C

At the beginning of this chapter, I asked you to visualize an event in the distant future. During this moment, you were gazing across a far distance—so you were more likely to construe the world in broad terms. Therefore, you were more likely to construe an event, like eating, in broad terms (getting nutrition), rather than narrow terms (chewing and swallowing).

You also prefer information that matches your construal level. For example, every product has a structure that I call a *benefits hierarchy*, in which technical specifications are positioned at the bottom, while higher benefits are positioned at the top (see Figure 6D). Salespeople could persuade customers by emphasizing information that matches their construal level. Customers who are buying a camera for the distant future will prefer high-level benefits (e.g., capture memories). But if they are buying a camera right now, they will prefer concrete details (e.g., 5-year warranty).

To recap: You focus on the gist of something while far away, but you focus on the details while nearby.

The revere happens, too: When you see the details of something very clearly, you assume that you are closer. If you see "Boston"

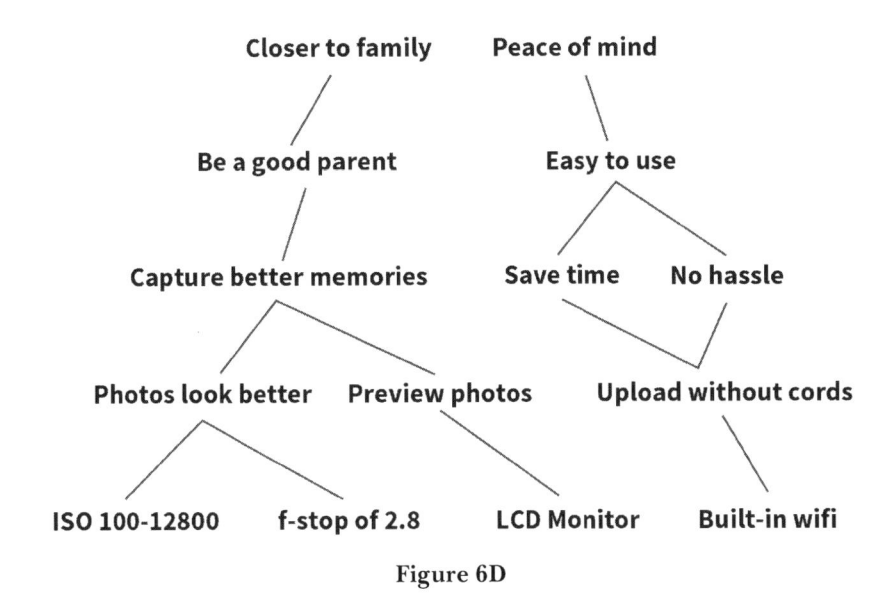

```
        Closer to family      Peace of mind
              /                      \
       Be a good parent         Easy to use
            /                   /         \
    Capture better memories  Save time   No hassle
          /        \           \          /
  Photos look better  Preview photos   Upload without cords
      /      \              \                  \
ISO 100-12800  f-stop of 2.8  LCD Monitor    Built-in wifi
```

Figure 6D

written in a clear font, you believe that you are geographically closer (Alter & Oppenheimer, 2008). Same with time: "January 15" should seem closer in a font that is easy to read.

Fluent stimuli seem closer, so you focus on the details of these stimuli (as if you are spatially closer). In one study, participants read "New York" in different fonts. When the font was easy to read, people wrote concrete details (e.g., *New York is a large city with five boroughs*; Alter & Oppenheimer, 2008).

Disfluency has the opposite effect. For example, think of a happy memory. Got one? Now, imagine this event while frowning. Feels weird, right? Okay, now answer this question: Is a camel a vehicle?

You just felt a weird sensation called *mind-body dissonance*. To make sense of this confusion, you needed to step back—metaphorically— by focusing on the broad gist. During this moment, you were more likely to conceptualize a camel as a vehicle because you were viewing the world in broad terms (Huang & Galinsky, 2011).

Verticality

Did you notice that we describe construals in terms of location? Why do we refer to a "high-level" perspective of something?

Once again, you can blame the sensory world: Higher viewpoints help you see the overall landscape. Whenever you feel higher up in space, you tend to focus on the broad gist of information. For example, people who walked up a staircase categorized objects more broadly (Slepian, Masicampo, & Ambady, 2015).

Same with eye gaze: Looking down triggers a low construal, while looking up triggers a high construal (Van Kerckhove, Geuens, & Vermeir, 2014). You're probably looking down at this book, right? I see you. Your visual perspective—*right now*—is dictating your conceptual perspective. While looking down, you might be fixated on the specific details that I'm describing (like this one). Perhaps you'd conceptualize the information more broadly if you lifted the book to eye level, especially if you held the book farther away.

Desirability vs. Feasibility

Finally, DISTANCE has a strong relationship with desirability and feasibility. Distant events make you focus on the "why" of behavior, whereas close events make you focus on the "how."

Consider a skip trip:

> . . . when one plans a ski vacation for a rare free weekend a few months down the line, the decision is made with thoughts of swooshing down powdery ski slopes and, perhaps, sipping hot cocoa at night in the lodge. However, on the morning of the trip, one may instead find him or herself bemoaning a 5-hr drive in the snow, the hassle of putting on all the equipment, and the long

wait to buy a lift ticket (Sagristano, Trope, & Liberman, 2002, p. 364).

While using a dating app, you might *seem* interested in somebody, yet once you connect, you often hesitate with second thoughts. What happened to the initial attraction? In the beginning stages, you are focused on the "why" of this person. Once you start talking, however, you move toward the "how" of this person. Suddenly your life and schedule seem more chaotic.

Want to schedule a meeting with someone? But don't have a good reason? Schedule the meeting sooner rather than later:

> . . . an individual deciding whether to schedule an appointment in the next several hours may consider if, where, and when he or she can do it, whereas an individual deciding whether to schedule an appointment in the distant future may consider whether doing so would be worthwhile, valuable, or desirable (Milkman, Akinola, & Chugh, 2012, p. 1).

Researchers emailed 6,548 professors pretending to be a doctoral candidate who asked for a meeting either that day or one week later. For same-day requests, professors showed no bias. For appointments next week, however, professors were 26% more likely to book appointments with white men (Milkman, Akinola, & Chugh, 2012).

I'll expand on those implicit biases later in the book.

Summary

You acquired a primitive concept of DISTANCE. Today, all types of distance—time, numbers, social—are built from the same sensory framework of spatial distance.

DISTANCE also activates a construal level:

- ▶ You focus on the gist of distant stimuli
- ▶ You focus on the details of nearby stimuli

Imagine that you see a suitcase. If the suitcase is far away, you might focus on broad concepts (e.g., travel in general, past vacation, next destination). If the suitcase is near you, however, you might focus on specific details (e.g., type of material, where to store it, when to pack).

You also prefer stimuli that match a construal level. Suppose that a city wants to reduce littering on the subway. The timing of this policy could influence the voting behavior: Residents might vote for more trash cans (specific solution) in the near future, but they might vote for an eating ban (broad solution) in the distant future (Rim, Hansen, & Trope, 2013).

Other examples:

- ▶ **Zoom Levels.** While using a computer, you can adjust the zoom level to activate a helpful construal level. Skimming a document? Zoom outward so that you broaden your focus toward the gist. Editing a document? Zoom inward so that you narrow your focus toward the details.
- ▶ **Zoomed Images.** Suppose that you're selling tickets to a future event, such as a virtual conference. Is the event far away? Customers will prefer a zoomed-out version of the ticket because they possess a high construal. Is the event happening soon? They will prefer a zoomed-in version of the ticket (see Ho, Kuan, & Chau, 2015).
- ▶ **Grayscale Images.** Your mental imagery is less colorful for events in the distant future. Therefore, you prefer black-and-white images for distant events. When participants evaluated a fundraiser for orcas in the distant future, they donated more

money when the orcas were in black and white (Lee et al., 2016).
- **Waiting Lines.** Did you ever notice that many waiting lines (e.g., banks, airports, Disney Land) use zig zag patterns to construct their lines? It's a clever ploy. People are now spatially closer to their destination, and they confuse this proximity for a shorter wait time.
- **Serial Numbers.** You prefer products with lower serial numbers (e.g., No. 3/100) because you feel closer to the origin or creator (Smith, Newman, & Dhar, 2015).

At this point, babies start noticing movement across spatial distances. The next chapter will set that plan in MOTION.

7

Motion

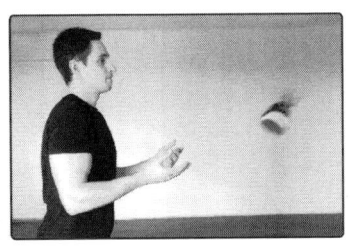

Left Toss | Right Toss

Figure 7A

IN FIGURE 7A, I'm tossing my shoe outward. Both images are mirrored versions of the same event, yet . . . does one throw seem stronger? Or . . . serendipitous? In this chapter, I'll explain why the rightward toss seems more familiar. Try and solve it.

Types of Motion

Objects are moving in the sensory world, so you insert this MOTION into abstract concepts. Compare these sentences:

- **Literal.** She climbed up the hill.
- **Abstract.** She succeeded in the company.

While reading both sentences, people were quicker to press a button that was located upward (Santana & De Vega, 2011). You understood both sentences by simulating upward motion.

Motion can also feel alive, as if the source is a human or animal. You identify *animate motion* in different ways:

- ▶ **Self-Motion.** Movement that isn't caused by another force.
- ▶ **Intentional Motion.** Purposeful motion, such as an object "chasing" another object.
- ▶ **Biological Motion.** Movement from a living thing.

When babies play with a mobile above their crib, we think: *Aw, that baby enjoys the toy.* Babies, however, don't realize they're playing with a toy. This interaction resembles their interactions with people, so they believe the mobile is a living thing (Mandler, 1992).

Refer to Figure 7B for a list of different motions.

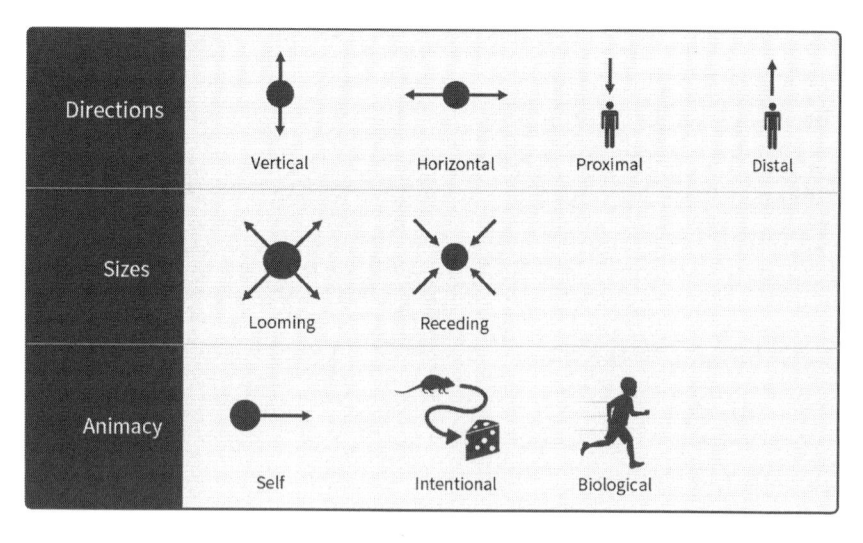

Directions	Vertical	Horizontal	Proximal	Distal
Sizes	Looming	Receding		
Animacy	Self	Intentional	Biological	

Figure 7B

Speed

Motion also has SPEED, which is another flexible idea that you insert into many domains. When people were asked to recite a speech very quickly, they completed a follow-up survey faster because they activated a primitive idea of SPEED (Shen, Wyer Jr, & Cai, 2012).

In Chapter 2, you learned that thinking of a car activates a barrage of sensory associations—such as visual, auditory, and motor cues. Speed is another cue that becomes activated.

What exactly becomes activated? Well, what would a fast object look like? It would probably lean forward, right? Greater leaning would indicate faster speed. Sure enough, people were quicker to read the word "cheetah" in italicized text because the forward leaning of the text matched their simulation of a cheetah (Walker, 2015).

Do you promote a "fast" service? Italicize your logo so that the name is leaning forward, as if it's moving fast.

Another question: Why does motion (and italicization) feel natural moving toward the right? Imagine a cheetah running toward the left. Feels weird, doesn't it? But why? Cheetahs can move in any direction. Why does leftward motion feel awkward? And for that matter, why does Super Mario move toward the right? And why does shoe tossing seem stronger toward the right?

You can blame *canonical motion*.

Canonical Motion

As you read this paragraph, this sentence will appear to the left of the next sentence. Now this sentence appears on the right, but it will soon appear to the left of the next sentence. And repeat this process.

Anyone who reads from left to right will encounter this pervasive

74   The Tangled Mind

Figure 7C

pattern in which prototypical motion—*canonical motion*—moves from left to right. That's why you have trouble visualizing a cheetah moving toward the left. And that's why shoe tossing seems stronger toward the right.

Oh, and memorize the image in Figure 7C. We'll revisit this image soon.

Simulated Motion

Somebody dies in a car accident every 27 seconds (ASIRT, 2018). Many accidents occur because people aren't paying attention, and they miss road signs. Well, most warning signs don't portray motion . . . we should change that.

Motion captures your attention automatically. Evidence confirms that warning signs are more likely to capture your attention when they depict a moving object (see Figure 7D; Cian, Krishna, & Elder, 2015).

You can find a similar effect in Figure 7E. Take a guess: Would this object fall or stay upright? Hold this guess.

Figure 7D

Figure 7E

For now, let's try another exercise. Figure 7F has a group of Λ's on the left. Can you find the hidden V? Once you find it, look for the hidden Λ in the other group.

All else equal, you were more likely to find the V faster (Larson, Aronoff, & Stearns, 2007). Why? Researchers blame threat detection. Back in the day, our ancestors needed to detect threats (e.g., angry people) very quickly; otherwise, whoops . . . we died.

How is a V-shape involved with threat? Some researchers argue that we detect anger through the "V" that appears in the downward formation of the eyebrows (Lundqvist, Esteves, & Ohman, 1999).

However, I disagree with that conclusion. I think a stronger

Find **V**   Find **Λ**

ΛΛΛΛΛΛΛ   VVVVVVV
ΛΛΛΛΛΛVΛ   VVVVVVV
ΛΛΛΛΛΛΛ   VΛVVVVV
ΛΛΛΛΛΛΛ   VVVVVVV

Figure 7F

Equilibrium Point

Example from earlier

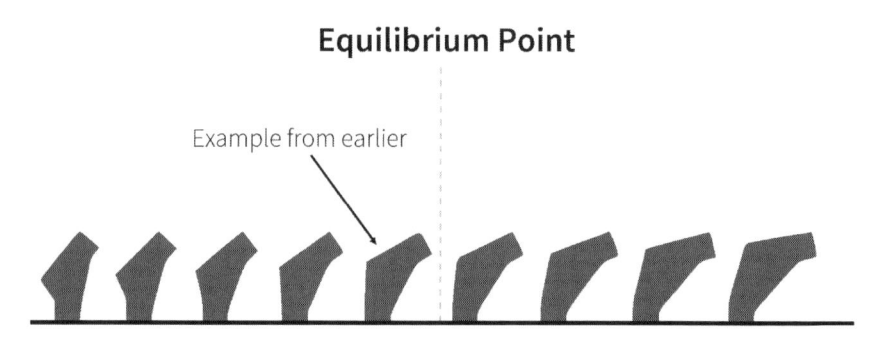

Figure 7G

explanation is *motion capacity*. You learned that motion captures your attention automatically. Well, a V-shape has a greater capacity for motion because you can simulate a swaying motion from side to side. You can't simulate that motion in a Λ-shape because it has legs. It would remain stable.

Remember the asymmetrical shape from earlier? Most people guess that it would fall. Did you?

In actuality, the shape would remain upright. I believe that the same effect is occurring—you are biased to see a premature tipping point because you are simulating the forward motion of this object (see Figure 7G).

In sum, you simulate motion. If you see an object that is moving, or has the capacity to move, you simulate this motion.

As you'll see next, you exhibit this effect in abstract domains.

Abstract Motion

You learned that DISTANCE is the primitive framework in all forms of distance (e.g., time, numbers, social).

Likewise, MOTION is the framework in all abstract forms of motion—so these abstract ideas inherit the sensory nature of motion.

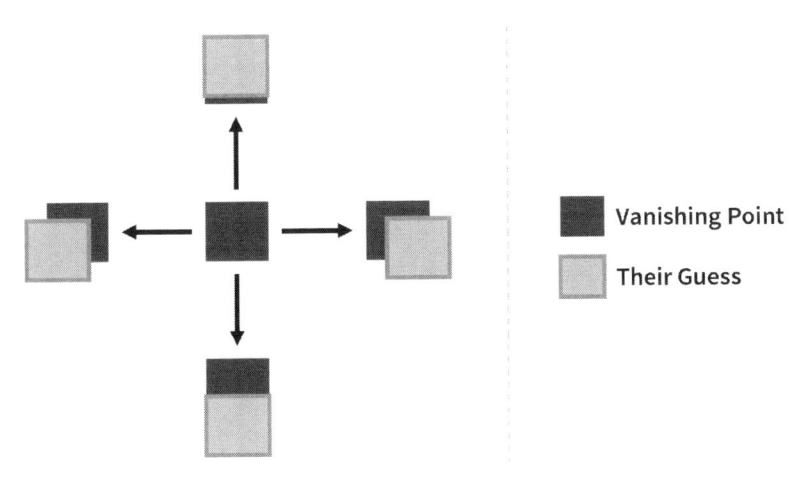

Figure 7H

For example, upward motion is difficult in the sensory world because of gravity. Therefore, you insert this discrepancy into abstract motion, such as traveling north. People believe that northern routes are more difficult because they paint this imagery with sensory motion and gravity (Brunyé, Mahoney, Gardony, & Taylor, 2010). You even believe that people lose more calories when they travel north, and you expect moving companies to charge higher shipping fees (Nelson & Simmons, 2009).

All forms of motion will adhere to the physical laws of motion, such as momentum. In some studies, people watch a moving box that suddenly vanishes on screen, and they mark the point at which they believe it disappeared. Figure 7H shows a typical response pattern (see Hubbard, 2005). People exhibit forward displacements in all directions. You can also see the influence of gravity: Upward motion has the shortest displacements, while downward motion has the farthest displacements.

Don't flip back, but do you remember the picture where I dropped my shoe? Figure 7I shows two variations. Can you choose the correct picture from earlier?

A   **B**

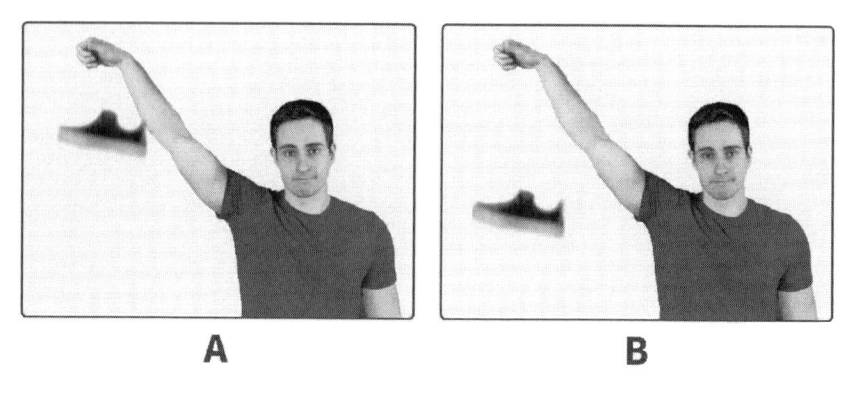

Figure 7I

Did you choose B? Sadly, nope. That's incorrect.

If you chose A, you might be feeling smart, huh? Well . . . neither picture appeared. Both pictures occurred a few milliseconds before (Variation A) and after (Variation B). I wanted to illustrate that you were more likely to choose Variation B because you simulated the downward momentum of the shoe.

As you'll see next, you insert this sensory momentum into abstract forms of motion.

Abstract Momentum

You conceptualize many abstract ideas with motion, imputing these ideas with the law of momentum. Stock prices going up? You expect them to keep rising. Win a few hands of poker? You expect to keep winning. Sports team catching up? You expect them to win the game.

Here are other examples.

Operational Momentum. You find a puzzling bias in mathematics: People overestimate addition problems, while they underestimate subtraction problems (Hubbard, 2015). For example, the correct answer to 8 + 8 is 16, yet—if answering quickly—you believe that 20 is more accurate than 13. Conversely, the correct answer to 16 − 8 is 8, yet you believe that 6 is more accurate than 10.

On the surface, this effect seems unrelated to motion—yet, when you look closely, you can spot momentum lurking underneath.

Think about it. What happens in addition problems? An entity is getting larger, right? Therefore, the final entity seems even larger because of the expanding momentum. In subtraction, however, an entity is getting smaller, so you perceive the final entity to be even smaller because of the shrinking momentum.

These simulations are so ingrained that 9-month-old children experience this bias while adding and subtracting blocks (McCrink & Wynn, 2009).

Temporal Momentum. Suppose that you're driving from Point A to Point B. When you reach the exact midpoint, you'll feel closer to Point B because you simulate your forward momentum. On subways, westbound riders feel closer to westbound stations, yet eastbound riders feel closer to eastbound stations (Maglio & Polman, 2014).

This effect can explain a puzzling effect with time. Visualize yourself one week in the past. Are you thinking? Good. Come back to the present. Now, visualize yourself one week in the future.

Both days, past and future, are equidistant from your present self by one week. Yet, does one side feel closer? Most people say the future because they paint this imagery with sensory motion, displacing their forward momentum through time (see Figure 7J; Caruso, Van Boven, Chin, & Ward, 2013).

Behavioral Momentum. People strive for consistency in their behavior. For example, you can persuade people by starting with small requests (Cialdini, 2007). Most households rejected a request to place a large and ugly sign in their front yard, yet researchers increased

80   THE TANGLED MIND

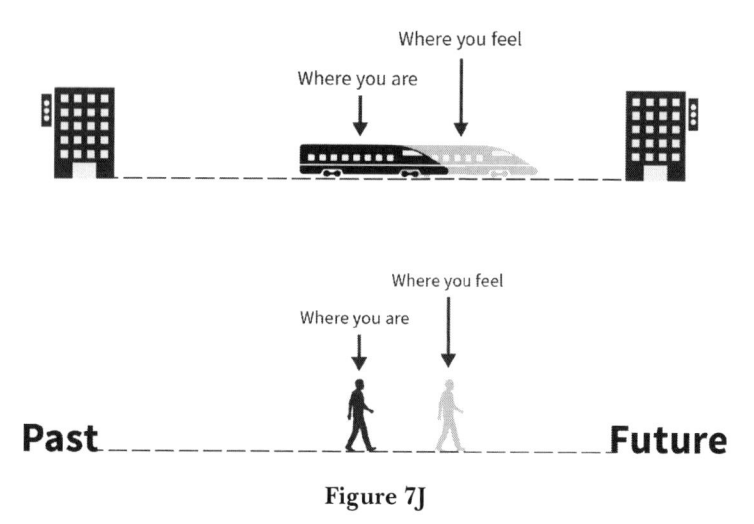

Figure 7J

that compliance by starting with a small request: Install a small sign. Most households complied with this small request, and they became *more* likely to install the large sign at a later date.

Perhaps a small request activates a feeling of motion. And, if you are moving—albeit metaphorically—you feel compelled to stay in motion because of the law of momentum.

Causal Motion

Sensory motion fueled your understanding of causality: Moving objects "cause" other objects to move.

Causality is built with motion—and, as you learned, you simulate motion. Therefore, when you see a cause, you simulate the effect because this outcome is the forward motion.

For example, while browsing Netflix, I noticed a thumbnail for *Jeopardy!* that showed a board with missing questions. This snapshot was devoid of visual motion, but it still has causal motion (see Figure 7K). Just like a warning sign in which you simulate the downward

Cause
Full Board

Effect
Empty Board

Figure 7K

motion of a falling rock, this thumbnail captures your attention automatically because you simulate the forward motion: You simulate a board with even fewer questions remaining.

Symbols of Causality

You can influence the perception of causality by adjusting the symbols that depict causality. For instance:

- **Size of Causes.** Bigger stimuli cause bigger effects. Referees call more penalties on tall players because the force seems stronger (Van Quaquebeke & Giessner, 2010).
- **Location of Effects.** Causality originates from the left because of canonical motion. Suppose that Person A pushed Person B. On which side was Person A—left or right? Most people say the left (Chatterjee, Southwood, & Basilico, 1999).
- **Distance Between Cause and Effect.** Causes seem stronger near the effect. Product claims seem more truthful on packaging (vs. advertisements) because they are closer to the product (Fajardo & Townsend, 2016).

To recap, you attribute causality to stimuli that are bigger, on the left, and closer to the effect. Let's apply these insights with a real-world implication: false confessions.

Why would anybody confess to a crime they didn't commit? The mere idea seems absurd . . . and juries think so too. When they see a confession—*any* confession—they assume that it's true.

Yet, in reality, many innocent people confess because of dirty interrogation tactics (e.g., intimidation, exhaustion, lies about evidence). Our justice system is sending innocent people to jail, and we need to fix this problem.

Perhaps these verdicts are influenced by sensory variables in confession videos. Bigger stimuli cause bigger effects, right? Well, camera perspectives can distort these visual sizes. Compare these perspectives inside a video frame:

- Suspect only
- Interrogator only
- Both suspect and interrogator

Many interrogations are filmed by depicting the suspect only, which seems logical:

> . . . confessions are customarily recorded with the camera lens directed at the suspect . . . a careful examination of not only suspects' words but also their less conspicuous actions or expressions, will ultimately reveal the truth of the matter (Lassiter et al., 2005, p. 34).

But these perspectives are problematic. With more visual emphasis on suspects, juries attribute more causality to the suspect. These camera angles lead to higher conviction rates because the suspect seems guiltier (Lassiter, Diamond, Schmidt, & Elek, 2007).

So, why not film suspects *and* interrogators inside the same frame?

That's a great idea . . . but we run into another issue. The position of the suspect—left or right—could influence the verdict as well. Causality originates from the left, so perhaps suspects seem guiltier on the left. Or perhaps juries attribute more causality to interrogators on the left, as if they forcefully extracted the confession.

And perhaps we should consider the size of the interrogation table. Smaller tables will bring interrogators closer to the suspect, a location in which they will seem more instrumental in extracting the confession. Confessions might seem more sincere when interrogators are farther away.

We usually think of the legal system to be an infallible judgment of truth. But it's not. You just learned that the visual traits of a table could determine somebody's life. We need to understand these biases so that we can remove them from important decisions in society.

Mind Reading

Think of a number between one and ten. Got something?

Whenever you encounter that question, you assume that you have ten possible choices with an equal chance of selection. In actuality, though, the distribution is far less random. Figure 7L shows the selections that people actually choose (albeit a small sample of 223 people). People choose 7 most often (followed by 8, 5, 9, and 3). In this section, I'll explain why.

First, some backstory. In college, I performed as a "mind reader" who accomplished feats using psychology. While opening my show, I asked the audience to think of a number between one and ten (and I subtly motioned the number seven with my hand gestures). When I asked who was thinking of seven, most people raised their hands.

At the time, I naïvely believed that my gesture was, indeed, influencing people to choose that number. I ran a study for my senior

Think of a Number Between 1 and 10
n = 223

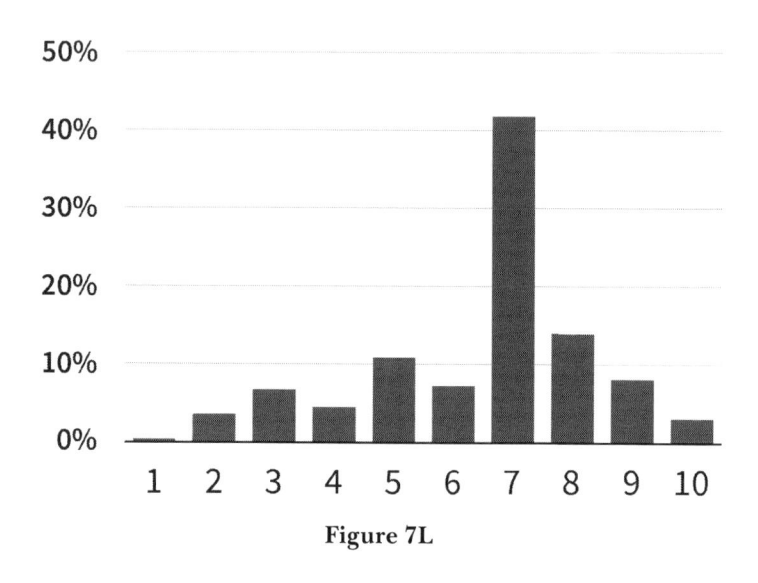

Figure 7L

thesis to confirm this effect. I asked one group to watch a video with my instructions, while I asked another group to listen to the audio. That way, both groups received the same instructions, yet only one group received the visual cue. The results crushed my naiveté... the gesture did nothing. Most people chose seven regardless of any cue.

Why did people choose seven most often? In my study, participants gave various reasons: My lucky number is 7 ... My birthday is December 7th ... My birthday is in July.

But remember, reasons are often meaningless. Many people generate those reasons *after* choosing a number.

Instead, I suspect that most people choose seven because they immediately think of five, and then adjust upward (usually landing on seven). I'm arguing that case for three reasons.

- ▶ **Central Fixation Bias.** Most people immediately think of 5 because of the central fixation bias. When you conceptualize the numbers 1 through 10, you transform those numbers into a horizontal spectrum (see Chapter 5 on Location). And, whenever you see a visual stimulus, you immediately look at the center. Here, the center is five. At this point, some people remain at this number—which explains why many people choose five—but other people will adjust upward for the following reasons.
- ▶ **Canonical Motion.** If people think 5 is too obvious, they'll adjust away from it. Based on canonical motion, they will adjust rightward—that's why most responses appear toward the right half of the distribution.
- ▶ **Handedness.** People might also adjust rightward if they are right-handed. You might see larger numbers (e.g., 8, 9) from people who rely more heavily on their right hand. Or you might see smaller numbers from left-handed people.

Summary

Babies are thrusted into a world with sensory motion, and they insert this idea into abstract concepts. Today, you conceptualize many intangible ideas—such as time or causality—with motion, as if these ideas are actually moving.

As a result, this imagery inherits the sensory nature of motion, including the law of momentum. For example, universities seemed more prestigious when their ranking ascended from 6th to 4th (vs. descended from 2nd to 4th). The final ranking—4th place—was the same, yet the ranking seemed superior after ascending because of a momentum effect (Maglio & Polman, 2016).

Other examples:

- **Resisting Momentum.** Resisting sensory momentum can make you resist abstract momentum. Researchers asked people to feel a sample of fur against the grain, and these people chose a less popular T-shirt. In addition, people who imagined walking against foot traffic were more likely to choose a less popular box of chocolates (Kwon & Adaval, 2017). Going against the flow in a sensory context can make you go against the flow in an abstract context.
- **Conversations.** Researchers digitally swapped the placement of two people in a conversation. In both versions, participants believed that the person on the left set the tempo and was better at conversing (Puccinelli, Tickle-Degnen, & Rosenthal, 2006). Perhaps, in late-night talk shows, the position of the guest influences your perception of the conversation.
- **Sales Presentations.** Delivering a presentation? Spokespeople were more persuasive toward the left of the product (Puccinelli, Tickle-Degnen, & Rosenthal, 2006).

So far, we've covered five primitives: SIZE, OBJECT, LOCATION, DISTANCE, and MOTION. Let's keep "moving" to the next primitive, which is slightly different. It's SHAPE.

8

Shape

IN A GROCERY STORE, customers who are holding a shopping basket (vs. pushing a cart) might be more tempted to buy ice cream. Any guess why? We'll see an answer later.

Affordances

While studying online behavior, I noticed that website visitors are sometimes more likely to click transparent buttons (see Figure 8A). That puzzled me. Wouldn't people prefer a colorful button that "pops" on a page? I'll give you time to brainstorm a reason.

For now, recall your tendency to simulate actions, such as grabbing the handle of a mug.

When do you perform these motor simulations? And what exactly do you simulate? The answer depends on the *affordance* of the object.

Certain objects allow (or "afford") interactions. Some doors have

Figure 8A

vertical bars, whereas other doors have horizontal plates. In both cases, the affordance is congruent with the motor action: Vertical handles match the grip of your hand, so you intuitively grip and pull; horizontal plates match the flat of your palm, so you intuitively press and push (Norman, 2013).

Motor simulations can distort your perception of objects. Hills seem steeper when you are wearing a heavy backpack because you exert more effort to imagine climbing: *Hmm, that hill seems difficult to climb. It must be steeper* (Bhalla & Proffitt, 1999).

When Do You Simulate Affordances?

You're surrounded by affordances, but you can't simulate *all* possible interactions at *all* times. So, when do you simulate an action?

You need to meet four conditions.

1. Affordance. Obviously, you need an affordance. Until now, we've been dealing with physical affordances. Small objects, like cherries, afford grasping with your thumb and finger (precision grip). Large objects, like apples, afford grasping with the palm of your hand (power grip).

Well, wait a second. Those affordances involve physical size, right? And you inserted physical size into numbers? If so, wouldn't numbers inherit these affordances? Interestingly, yes. People were quicker to perform precision grips while viewing small numbers, and they were quicker to perform power grips while viewing large numbers (Van Dijck, Fias, & Andres, 2015).

Also, many objects have *multiple* affordances. If you wanted, you could use a power grip to hold a cherry. You might look weird, but hey, no judgment.

So, perhaps we have *canonical affordances*: Grabbing a mug handle will be the most common interaction—and thus the interaction that you simulate (see Figure 8B).

Shape 89

Grabbing Handle

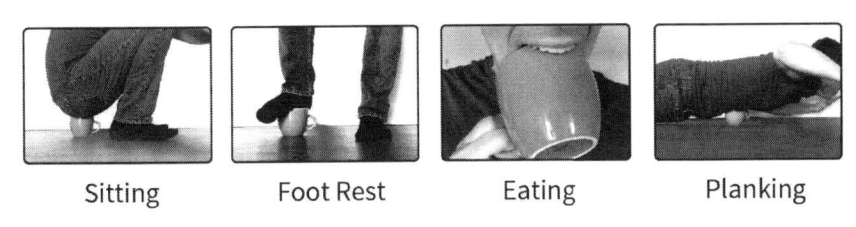

Sitting Foot Rest Eating Planking

Figure 8B. Grabbing a mug handle is the canonical affordance. Other interactions are less common.

2. Accessibility. Even if an affordance is present, you need to possess the physical ability to interact with it:

- **Reachable.** Objects need to be within your reach (see Witt & Proffitt, 2008).
- **Free Hands.** Even if an object lies within reach, your hands need to be free (Shen & Sengupta, 2012).
- **Barricades.** People are less excited by food behind a plexiglass barrier because they can't imagine grabbing the food (Bushong, King, Camerer, & Rangel, 2010).
- **Tools.** Objects seem closer (and more desirable) if you are holding a baton because it extends your potential reach (Witt, Proffitt, & Epstein, 2005).

Accessibility depends on *anticipated* accessibility. You merely need to see a baton on the table because you can imagine grabbing the tool and *then* grabbing the object (Witt & Proffitt, 2008).

3. Personal Ability. Your simulations are also influenced by your personal traits:

- **Age.** Older adults perceive hills to be steeper because they exert more effort to imagine climbing (Bhalla & Proffitt, 1999).
- **Weight.** Distances seem farther for people who are overweight or wearing artificial weight (Sugovic, Turk, & Witt, 2016; Lessard, Linkenauger, & Proffitt, 2009).
- **Energy.** Hills seem more climbable after consuming an energy drink (Schnall, Zadra, & Proffitt, 2010).
- **Training.** To a normal person, walls don't afford climbing. To a parkour expert, however, walls *are* climbable. Thus, parkour athletes perceive walls to be shorter because they can imagine climbing them (Taylor & Witt, 2010).
- **Disability.** Paraplegic patients had trouble following a sequence of lights that mimicked bodily motion because they couldn't simulate the depicted motion (Arrighi, Cartocci, & Burr, 2011).

4. Intention. Finally, you must *intend* to perform the interaction. Without intention, you don't simulate. Without simulation, you don't feel an emotion of ease or difficulty. And without ease or difficulty, you have nothing to distort your perception.

Yet a loophole exists: You simulate affordances that are congruent with an active simulation. In one study, people were quicker to find an orange when they were holding a spherical object, compared to a hockey puck (List, Iordanescu, Grabowecky, & Suzuki, 2014). You notice affordances that are congruent with your body state: When you're a hammer, everything looks like a nail.

Have you solved the mystery from the beginning of the chapter?

How could grasping actions—pushing a cart vs. holding a basket—influence the products you buy? While holding a basket, you might be tempted to buy frozen foods (e.g., ice cream) because the affordance of the basket handle matches the affordance of the freezer handle. You can simulate opening a freezer door more easily because your arm is activated from the basket.

But Nick . . . if people are holding a basket, wouldn't their hand be occupied? Wouldn't their simulation be weaker?

I considered that, too. But remember: Simulations depend on *anticipated* ability. Shoppers still have a free hand, so they can imagine moving the basket into their free hand. This effect should still hold, so to speak.

In addition, this pulling motion could strengthen your desire for the intended food. Perhaps you'd buy a larger tub of ice cream if the freezer door was heavier: *Hmm, I'm exerting a lot of effort to pull open this door. I must really want ice cream. I'll buy the larger tub.*

How to Facilitate a Simulation

People are more likely to buy products if they can imagine interacting with them (e.g., grabbing the handle). Therefore, you can strengthen desire for objects by strengthening these simulations. You just need to adjust one of the four conditions: affordance, accessibility, ability, or intention.

For example, you could push objects closer to people. Figure 8C shows a screenshot from my iPhone in which the payment window is near the bottom of the screen, a location that is closer to your fingers.

Coincidence? Maybe. But probably not.

This location helps you imagine pressing the payment button. And you confuse this fluency for a desire to push the button.

Earlier, I asked why people can be more likely to click transparent buttons on the computer. My guess? The empty space affords movement. You can imagine pressing your finger into this empty space,

92   The Tangled Mind

Figure 8C

simulating the sensory enclosure around your fingers. Researchers could test that argument by asking people to hold something while evaluating a transparent button—their preference for transparency should weaken if their hands are occupied.

You can apply these principles to any scenario. Struggling to find motivation to exercise? While sitting down, your heart rate and energy are low (and your simulation of exercising is weak). To overcome that hurdle, you could pace across the room. You'll increase your heart rate and energy—an aspect of personal ability—and your simulation of exercising will become stronger.

You can influence any decision by strengthening the vividness in which people imagine this behavior. Refer to my book *Imagine Reading This Book* for a plethora of other examples.

Roundness

SHAPE also has intrinsic qualities, such as roundness.

Evidence shows that humans prefer round objects (Bar & Neta, 2006). Why? Some researchers argue that sharpness seems threatening, but this effect remains a mystery because evidence shows that we also gravitate *toward* roundness:

> ... preference for curvature cannot derive entirely from an association of angles with threat ... Further studies are needed to clarify the nature of such a preference (Palumbo et al., 2015, p. 1).

My guess? Round objects afford grasping.

Your hands form a natural curve that encapsulates round objects very easily (and without pain). You prefer round shapes because you can imagine these motor interactions more easily.

Researchers could test that explanation by asking people to rate shapes while holding something. Round objects should seem better on the dominant side, but the preference should weaken if people are holding something.

Precision

In the sensory world, many objects have different shapes and contours—and this SHAPE becomes infused with abstract ideas. For example, we describe numbers in terms of roundness: Sharp numbers are precise, while round numbers lack precision.

Therefore, perhaps we learn numbers through SIZE *and* SHAPE. Research shows that house prices seem cheaper when they are precise (e.g., $395,425; Thomas, Simon, & Kadiyali, 2010). Perhaps we conceptualize these sharp numbers with a small and fine point.

If that explanation is true, it might clarify past findings. For example, round prices work better for emotional products, and sharp prices work better for rational products. Champagne was preferred with a round price ($40), yet a calculator was preferred with sharp prices ($39.72 and $40.29; Wadhwa & Zhang, 2014)

However, rational products are typically sharp (e.g., knife, pencil, toothpicks), while emotional products are soft and round (e.g., clothes, blanket, pillow). Perhaps the effect is caused by SHAPE. If so, you might prefer $20 for a hammer, which is rational *and* round.

Goal specificity could play a role, too. Consider the price of this book. You might prefer a sharp price, like $23.73, if you possess a specific goal to improve your business. Or you might prefer a round price, like $25, with a broad goal to expand your knowledge.

Containment

Babies start noticing another quality of SHAPE: Small objects can be inside large objects.

This sensory idea—*containment*—is more prevalent than you think. Read the following passage about a child's everyday experience. I counted ten containers. Can you spot them all?

> Take for example a child in a red dress who watches her mother put cookies into a jar . . . She reaches into the jar, reaches down into the cookies to find a particular cookie near the bottom, grasps the cookie (so that the cookie is now in her hand), and takes it out. She wraps the cookie in a napkin. She walks with the cookie through a door into another room, where she is picked up in her mother's arms and put into a high chair. She watches the mother pour milk into a glass. She then dunks her cookie into the milk (which is itself contained in the

glass), and she puts the cookie into her mouth (Dewell, 2005, pp. 371–372).

Babies learn that cookies can be IN a jar, and they extend this trait into abstract domains: You can be IN a good mood because you conceptualize this idea as a container.

Containers also help you learn OPEN and CLOSED. Jean Piaget was an influential researcher in child development. One day, he kept blinking to get his 9-month old daughter to blink, but it was like talking to a baby. Literally. Two months later, Jacqueline still didn't blink—but she performed other behaviors (e.g., opening and closing her hands and mouth). Blinking your eyes and opening your hands seem vastly different, but are they? Perhaps Jacqueline was learning a general concept of OPEN and CLOSED:

> . . . his children were expressing a concept of opening and closing. They had the right idea but couldn't locate the right part. The concept was abstract: opening and closing per se—not opening or closing a particular object but a more abstract representation of the act itself (Mandler, 2004, p. 32).

Abstract Boundaries

This section "contains" a powerful implication. Read the previous sentence again. You are conceptualizing this section of the book as a container.

Why does that matter? Because you are imputing this section with the sensory nature of containment. Right now, you aren't just inside a section of this book. You are inside physical container with rigid boundaries.

Don't believe me? Let's see this effect in different scenarios.

Geography. Every year, people die from natural disasters because they don't evacuate. They falsely believe their area will be safe:

> When people hear that an earthquake has occurred in Nevada, they categorize it as an event of Nevada that is more likely to spread to other locations within Nevada than to equidistant locations outside Nevada (Mishra & Mishra, 2010, p. 2).

Geographic regions are abstract, so you conceptualize this idea with CONTAINMENT. You impute these abstract boundaries with *physical* boundaries. In one study, potential home buyers compared two homes that were located 200 miles from an earthquake zone. If the home was located in a different state, they were 272% more likely to buy it even though it was the same distance away (Mishra & Mishra, 2010).

Time. Time has various categories—days, weeks, months, years. You conceptualize these ideas as sensory containers. Suppose that you need to complete a project within 5 days. Your motivation will be stronger if this deadline falls inside the current month (see Figure 8D). Deadlines seem less urgent "next month" because this date falls inside a separate container (Tu & Soman, 2014).

Since you are moving through time, you also conceptualize obstacles as sensory blockages:

> Harry got OVER his divorce. She's trying to get AROUND the regulations. He went THROUGH the trial. We ran INTO a brick wall (Lakoff & Johnson, 1999, p. 189).

Numbers. Numbers have physical boundaries, too. In any given day, stock prices are more likely to rise if they begin slightly above a

Shape

[Figure 8D: Two rectangular containers. Left container labeled "April" contains a human figure labeled "You". Right container labeled "May" contains a checkered flag labeled "Deadline".]

Figure 8D

rounded threshold (e.g., $80.02). Investors conceptualize the round number (e.g., $80) as a physical boundary, as if the stock price can't fall below that barrier (Johnson, Johnson, & Shanthikumar, 2007). Conversely, stock prices tend to decrease if they start below a rounded threshold (e.g., $79.98).

You also tend to group lists into sections (e.g., top 5, top 10). These sections feel like physical containers, so you ascribe more importance to items that cross these boundaries. Students are more likely to apply to universities that cross a rounded threshold in a popular ranking—moving from #11 to #10 is more impactful than moving from #10 to #9 (even though the latter change is better; Isaac & Schindler, 2013).

Communication

Whenever you speak to somebody, you encounter two primitives: MOTION and CONTAINMENT. You insert meaning into words, and you send this container to the recipient. You see these metaphors in language:

> I'll GIVE you an example to illustrate how we PACK meaning INTO words to get our point ACROSS.

Sometimes words can be EMPTY and meaningless, but sometimes they can be FILLED with emotion. Occasionally, the meaning can be so powerful that it MOVES us. Am I GETTING THROUGH?

Perhaps this metaphor can explain a unique effect in persuasion: If you realize that somebody is trying to influence your behavior, you become *less* likely to comply (Friestad & Wright, 1994).

Think about this imagery. When people speak to you, they are sending you a container of their words. Now, what happens when you feel a heavy force coming toward you? You need to push forward, right? If you walk against the wind, you need to push forward so that you don't fall backward. Therefore, if somebody is being forceful or pushy, you conceptualize this behavior as a heavy force coming toward you. Perhaps you return a forceful demeanor because you feel compelled to remain balanced.

Confinement

Containers also possess density, which is similar to a Korean term—*kkita*—that depicts the *tightness* of containment: A ring on your finger has tight containment, while a ring inside a box has loose containment (McDonnough, Choi, & Mandler, 2003).

Right now, you're located inside a container, such as the current room, and you are feeling a sensation of density. You might be in a spacious room, but sometimes you feel confined.

More importantly, how do you react when you feel spatially confined? You feel an urge to move, right? You want to reclaim your freedom. And that's exactly what happens, metaphorically.

Researchers created a store aisle that was wide or narrow, and they asked people to perform various tasks. In the narrow aisle, participants reclaimed their freedom by choosing uncommon options: They chose a greater variety of chocolates, and they donated to lesser-known charities (Levav & Zhu, 2009; Meyers-Levy & Zhu, 2007).

In that example, people felt confined from the boundaries. Sometimes, though, you feel confined because many objects are inside a container. Yet the same effect occurs. Based on 170 million store transactions, customers buy a wider variety of brands and flavors when more people are inside a store (Levav & Zhu, 2009).

Also, in those examples, your entire body felt confined—but *any* confinement should trigger this effect. Coincidentally, I just hopped out of the shower (a container). Being a single bachelor, and being totally fixated on writing this book, I've been neglecting other duties in my life, like laundry. I just dug into the deep recesses of my wardrobe (another container) to find a pair of socks (a third container) that I haven't worn in years. *And oh man they're tight.* Your entire body doesn't need to feel confined; any feeling of confinement should activate the same implications. While wearing tight socks, I'm more likely to behave in ways that I normally wouldn't because I'm trying to escape my confinement. Perhaps that's why I'm describing an anecdotal example instead of my usual academic rant.

Protection

Density feels confining, but boundaries feel protecting. Researchers blared emergency sirens while people solved math problems. Afterward, people who couldn't control the volume were more likely to choose a postcard with black borders, as if these boundaries could metaphorically protect them from the noise (Cutright, 2011).

In a follow-up study, people were more likely to buy a risky product when the logo had a visual border (see Figure 8E).

Perhaps most interesting, this effect is weaker in religious people: Atheists and agnostics exhibit a stronger preference for boundaries because they need to seek protection from an external source (Kay, Gaucher, McGregor, & Nash, 2010).

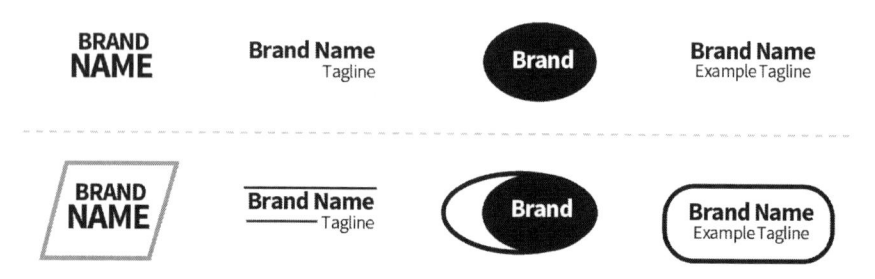

Figure 8E

Emotional Closure

When you stop feeling emotional reactions toward an event, you've reached *closure*. And again, this metaphor is no coincidence.

Emotional closure can be activated with physical closure. When people wrote about a regretful decision, they felt less negative emotion if they sealed their description inside an envelope (Li, Wei, & Soman, 2010). This enclosure blocked the negative emotions.

Going through a breakup? You can feel better by deleting past messages from your phone.

Summary

You are surrounded by objects with different shapes, and you inserted these SHAPES into abstract ideas.

For example, a natural disaster seems less likely to move from Maine to New Hampshire because you conceptualize these regions as containers, as if the state boundaries will block the disaster. However, if you conceptualize that region as New England, which contains Maine and New Hampshire, both states seem equally vulnerable because they fall inside the same container.

Shape **101**

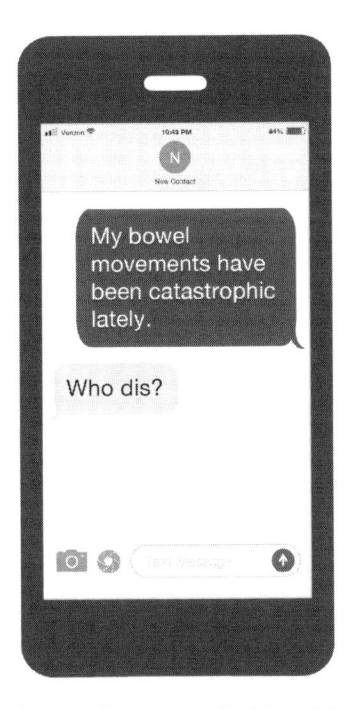

Figure 8F. Send an embarrassing text? You'll feel better by deleting that message from your phone.

You also evaluate objects more favorably when you can imagine interacting with these objects.

Other examples:

- ▶ **Visual Boundaries.** People believed that a radioactive spill would be less likely to cross state boundaries if the lines on a map were visually darker (Mishra & Mishra, 2010).
- ▶ **Product Numbers.** Many businesses update their products to new versions. Although a change from Version 3.4 to Version 4 is smaller than a change from Version 3 to Version 4, the first change seems larger because it crosses a categorical threshold

from decimal to integer (Shoham, Moldovan, & Steinhart, 2018).
- **Test Scores.** Students are more likely to retake the SAT if they scored just below a rounded threshold (e.g., 1190; Pope & Simonsohn, 2011).
- **Leave Cliffhangers.** You will perceive the end of this chapter as a physical boundary. And thus, your motion through this book will hit a wall (and you will be less likely to continue). I should probably leave an enticing cliffhanger to keep you going. That's why I left a fun exercise in the beginning of the next chapter.

9

Orientation

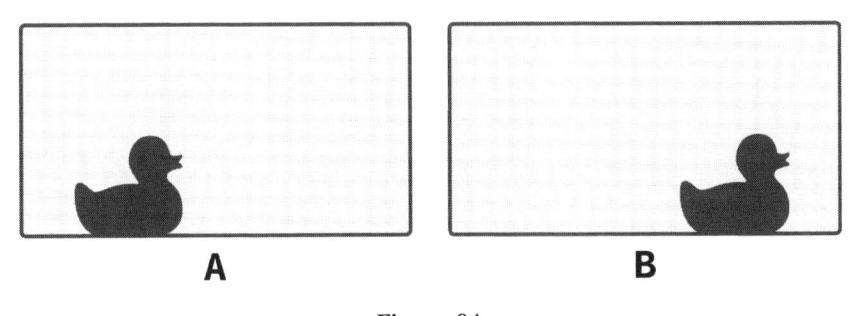

Figure 9A

IN FIGURE 9A, a duck is facing right. Yet, does one variation feel better? Is it Version A? But why? Try to articulate the reason before we reach the explanation.

Babies discover that objects have shapes, and they start noticing that these objects possess an ORIENTATION:

- ▶ **Angle.** Objects can be tilted or straight.
- ▶ **Direction.** Objects can face a particular direction.
- ▶ **Perspective.** Objects can look differently depending on your own spatial position.

You impute those traits into various ideas, as you'll see next.

Canonical Orientation

In Figure 9B, you see two photos of myself. If you've never seen my face, then you probably don't see a meaningful difference. Yet I prefer Variation A, while my friends and family prefer Variation B. In fact, Variation B makes me cringe. I prefer Variation A because that depiction is the mirrored version of myself that I see every day. Variation B is the frontal depiction that my friends and family see. In both cases, you prefer the depiction that matches the typical orientation—the *canonical orientation*.

Perhaps you'd find this effect in the vertical spectrum, too. Men instinctively take more selfies by holding their camera at a lower angle, creating an upward view of their face (Sedgewick, Flath, & Elias, 2017). Perhaps women also prefer these views. Women are typically shorter than men, so this upward view is their canonical perspective of men. It feels more familiar. If true, the data might reverse for short men (and tall women) whose canonical perspectives are reversed.

You also find *canonical directions*. If you conceptualize your

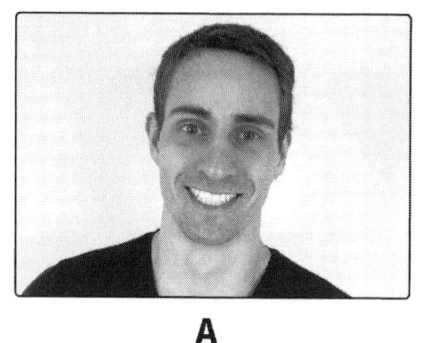

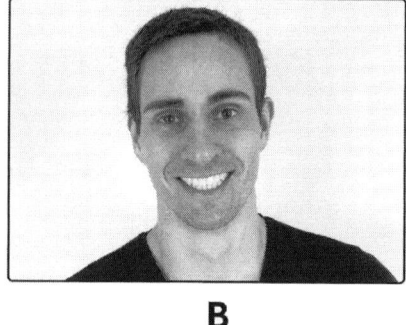

Figure 9B

yearbook as a way to reflect on past memories, you might prefer portraits that face the left; but if you conceptualize the yearbook as a way to look toward the future, you might prefer portraits that face the right.

These directions could also influence facial expressions. While posing for a picture, people were asked to show emotion—and they showed the left side of their face. However, when asked to suppress emotion, they showed the right side (Nicholls, Clode, Wood, & Wood, 1999). Based on the evidence, researchers concluded that left cheeks are emotional, while right cheeks are rational. But I disagree. Can you spot another explanation?

My guess: *Emotion has magnitude.*

When asked to show a large magnitude of emotion, people look right (thereby showing their left cheek) because they conceptualize larger magnitudes toward the right. So, it's not the cheek, *per se*. The real culprit is the location of size.

Attention Orientation

Before humans invented language, our ancestors relied on eye gaze to decipher information. Eye gaze became "hard-wired" into our brain (Emery, 2000). In fact, evolution may have facilitated this detection by increasing the saliency of our eyes:

> . . . the physical structure of the eye may have evolved in such a way that eye direction is particularly easy for our visual systems to perceive (Langton, Watt, & Bruce, 2000, p. 52).

Today, eye gaze captures your attention automatically—and these cues are additive: More cues, such as pointing, capture even more attention (Ariga & Watanabe, 2009).

Figure 9C

Suppose that somebody is choosing a photo for a Tinder profile. Which orientation in Figure 9C would strangers prefer?

All else equal, they'd prefer the version looking rightward.

In dating apps, users approve photos by swiping right. If they see a body that is oriented toward the right, they push their attention toward the right (and they feel a stronger urge to swipe right). The same effect occurs with images of shoes (Van Kerckhove & Pandelaere, 2018).

But wait . . . our ancestors didn't need to detect the direction of shoes. So, why do shoes guide our attention?

You can blame our tendency to *anthropomorphize*: We attribute human traits to inanimate objects. For example, human bodies have fronts and backs, which you attribute to other objects:

> . . . our bodies shape conceptual structure . . . We project fronts onto stationary objects without inherent fronts such as trees or rocks . . . If all beings on this planet were uniform stationary spheres floating in some medium and perceiving equally in all directions, they would have no concepts of front or back (Lakoff & Johnson, 1999, p. 34).

You derive the entire concept of FRONT and BACK from human bodies. And you extend this trait into inanimate objects. For example, digits can face a particular direction:

- **Left:** 1, 2, 3, 4, 7, 9
- **Center:** 0, 8
- **Right:** 5, 6

You recognize those directions because you perceive these digits as human bodies. It sounds weird, but if that's true, then these directions should guide your attention . . . and they do.

Consider two prices:

- $58
- $98

Let's ignore the 8-digit because it faces the center. However, the 9 faces left, while the 5 faces right. Those directions are influencing your perception of each price. With the right-facing digit in $58, you shift your focus toward the ones digit (in this case, 8). Because you focus on 8, you round that digit upward (and you perceive a final price closer to $60). However, with $98, you keep your focus leftward. Since you don't contemplate the ones digit, you round down (and you perceive a final price closer to $90; Coulter, 2007).

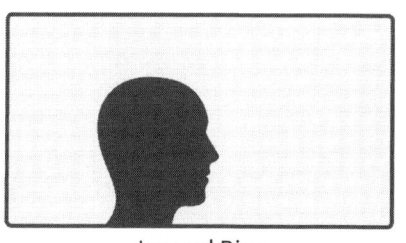

Center Bias   Inward Bias

Figure 9D

Visual Biases

You prefer objects in particular locations, depending on their orientation (Palmer, Gardner, & Wickens, 2008).

- ▶ **Center Bias.** You prefer front-facing objects in the center.
- ▶ **Inward Bias.** You prefer side-facing objects to face inward.
- ▶ **Rightward Bias.** You prefer side-facing objects to face right.

It sounds complicated in writing, but you can see all three biases with the face in Figure 9D. When it faces the back or front, you prefer the face to be in the center (center bias). However, when it turns to the side—left or right—you prefer the face to point inward toward the center (inward bias). Given a choice, though, you prefer a rightward direction (rightward bias).

You already know the answer to the rightward bias: You prefer stimuli to face the right because that's the direction of your canonical motion. But what causes the center and inward biases?

Some researchers suggest that you prefer the fronts of objects to be located near the center. When an object is facing toward you, the front is already in the center. But when an object faces the side, you need to horizontally nudge the object to position its front in the center (Leonhardt, Catlin, & Pirouz, 2015).

I also think the next two explanations are plausible.

Explanation 1: You Simulate Forward Motion

You simulate the motion of objects. Perhaps you experience an inward bias because you desire more space in front of objects, which allows you to simulate the forward motion.

Researchers could test this argument by comparing similar objects with varying levels of motion capacity: You might prefer more space in front of a roller skate (vs. shoe) because the skate has a greater capacity for motion.

Explanation 2: You Simulate the Affordance

Second, your simulations can be stationary or transportive.

Stationary simulation occurs when you simulate the interaction from your own spatial position. Suppose that you see an image of a computer. If that computer is facing the front, you prefer the computer in the center because you can simulate the interaction from your own position.

Transportive simulation occurs when you need to immerse yourself into a scene. Suppose that a computer is facing the right. In order to simulate the interaction, you need to immerse yourself in front of the computer. And what do you need for that immersion? Aha . . . space. You prefer objects that face inward because it adds extra space in front, allowing you to immerse yourself into this area.

In TV and film, directors often place actors toward the left-third or right-third of a frame; it's called the *rule of thirds* (see Figure 9E). Perhaps this framing creates extra space on the sides, allowing viewers to immerse themselves into the dialogue—as if they are the ones who are speaking to the character.

Some directors overgeneralize the rule of thirds by neglecting the center bias. Some online courses, for example, erroneously position

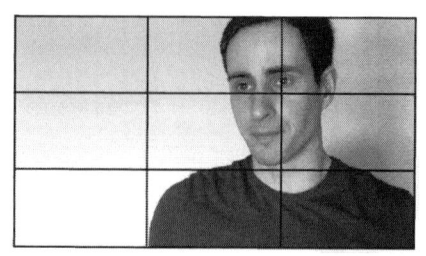

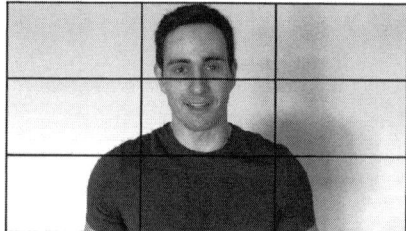

Figure 9E

instructors near the thirds of a frame. But instructors are usually facing the front, so viewers will prefer a center framing.

Summary

Humans face particular directions, and you attribute this trait to inanimate objects and ideas. Cars seem more powerful when they face toward you because you confuse the headlights for eyes (Schuldt, Konrath, & Schwarz, 2012).

Other examples:

- ▶ **UP and POWER.** Think of the blandest product imaginable. Maybe white rice? Well, even white rice seems more powerful when viewed with an upward angle (Van Rompay et al., 2012).
- ▶ **Willingness to Pay.** Researchers asked people to indicate how much they would pay for a product by sliding a marker to the right. When products were facing right, they pushed this amount further to the right (Van Kerckhove & Pandelaere, 2018).

10

Sound

Figure 10A

IN FIGURE 10A, which shape is bouba? And which is kiki?

Did you associate kiki with the pointy shape? And bouba with the round shape? Across the world, 95% of people choose these labels.

So far, you encountered these primitives: SIZE, OBJECT, LOCATION, DISTANCE, MOTION, SHAPE, and ORIENTATION.

Those primitives might be the most important concepts that shape your knowledge, partly because they occur in different modalities, such as sight and touch. Even if you are blind from birth, you still experience those primitives while navigating the world.

Nevertheless, other concepts—like SOUND—can mold your perception of the world, too. The word "bouba" just *feels* round, even though you can't articulate why.

Let's unravel these connections.

Components of Sound

Phonemes are the smallest unit of sound. The word "THE" has 3 letters, 1 syllable, and 2 phonemes (TH and UH). Every language pulls from a collection of roughly 44 phonemes (Harrington & Johnstone, 1987).

We can't trace all 44 phonemes to specific meanings, so we need to categorize these sounds—see a few examples below (Klink, 2000; Liberman & Mattingly, 1985).

- **Front Vowels.** Tongue in front (e.g., e, i, ē)
- **Back Vowels.** Tongue in back (e.g., ü, o, u)
- **Voiced Consonants.** Vibration occurs (e.g., b, d, g)
- **Voiceless Consonants.** No vibration (e.g., p, t, f)
- **Fricatives.** Air escapes your mouth (e.g., s, f, z)
- **Stops.** Air stops at your mouth (e.g., p, k, b)

No need to memorize them. We'll revisit them later.

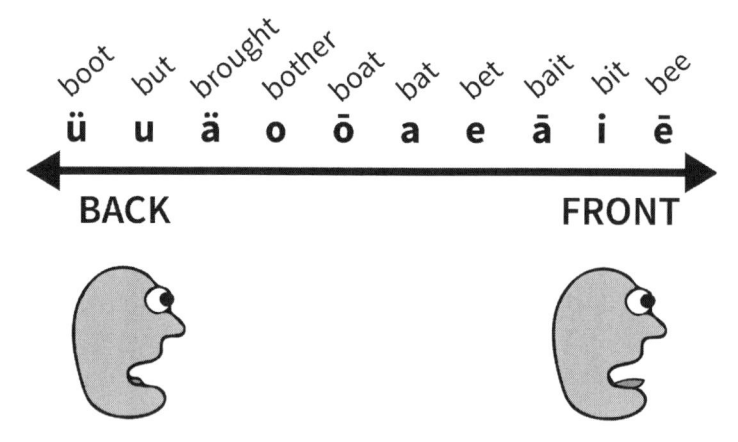

Figure 10B. Spectrum of vowels from back to front.

Canonical Sounds

When you think of a car, you simulate the sensory associations (e.g., visual, emotion, motor). Refer to Chapter 2 for more details.

You activate sounds, too. For example, researchers blared white noise while people named specific objects in images—turns out, the white noise slowed their ability to name objects with a typical sound, such as birds. They were unaffected while naming other objects, like buildings (Mulatti, Treccani, & Job, 2014).

A chirping sound is part of your concept for birds. White noise interferes with your ability to simulate this sound, so it interferes with your ability to identify a bird:

> For objects with a typical sound, the typical sound is part of the knowledge of that object and, therefore, the activation of the typical sound is part of the process of object identification (Mulatti et al., 2014, p. 1).

Some objects produce multiple sounds, so which sound do you simulate? You simulate the *canonical sound*.

This section will explain different types of canonical sounds.

Canonical Voices

Why do people hate the sound of their own voice? You can blame their canonical voice.

Whenever you speak, vibrations distort the sound of your voice—you hear a distorted voice, while other people hear your true voice. Over time, you believe that your distorted voice *is* your voice. Therefore, a recording of your voice sounds jarring because it's incongruent with the distorted voice that you expect to hear.

That same process can be detrimental in society. Accents are incongruent with the typical voice that you expect to hear, so you degrade your evaluation of the information. In one study, ambiguous trivia statements—Ants don't need sleep—seemed less truthful when spoken with a heavy accent: *Hmm, this statement doesn't feel right. It must be false.*

Sure, ants seem trivial—but that mechanism will influence your perception in other contexts, like medical advice from doctors. You might dismiss vital information about your health because of the irrelevant traits of your doctor's voice.

Canonical Phonemes

Let's try an exercise. Create a definition for this fictional word: *glon*. Keep it in mind.

Researchers used to believe that language was arbitrary. Besides an occasional onomatopoeia—*woof, bang, fizz*—sounds in language shared no symbolic meaning with their intended concept.

However, they now believe in *sound symbolism*: Certain sounds convey universal meaning across cultures and languages. For example, the bouba-kiki effect occurs in remote populations and 4-month-old babies (Bremner et al., 2013; Ozturk, Krehm, & Vouloumanos, 2013).

If symbolism is so great, why not symbolize *all* communication? Why not create a mapping so that every sound conveys clear meaning? That approach "sounds" great in theory, but it fails in practice:

> . . . if there were a close correspondence between form and meaning then the possibility of confusing the word for sheep with the one for cow is increased (e.g., if the two animals were referred to as feb and peb, respectively; Monaghan, Christiansen, & Fitneva, 2011, p. 327).

Figure 10C. The importance of arbitrariness in language.

We need both concepts: symbolism *and* arbitrariness. Symbolism reinforces the meaning (which creates vividness). Arbitrariness offers more letters (which creates efficiency).

But how do sounds acquire meaning in the first place? One source is blending. New words often emerge from past words.

- **Flurry:** flaw + hurry
- **Flare:** flame + glare
- **Slosh:** slop + slush
- **Splatter:** spatter + splash
- **Squiggle:** wriggle + squirm
- **Swipe:** sweep + wipe
- **Twiddle:** twist + fiddle

Pretty fantabulous, right?

Sounds inherit meaning because blending injects certain phonemes (e.g., SN) into similar words (e.g., snore, snout, snort, sniff, sneeze). When asked to create a term for removing black stuff from overdone toast, over 25% of people created a word with SK- because those sounds felt right (Magnus, 2001).

Remember your definition for glon? Even though the word was fictional, many people create a definition related to light or vision because we blended GL- with words in that domain:

> **GL:** *glimmer, glisten, glitter, gleam, glow, glare, glint*

Those examples involved semantic meaning, but phonemes can also inherit emotional meaning:

> . . . back vowels such as the [u] sound in dull or ugh are often found in English words expressing disgust or dislike (e.g., blunder, bung, bungle, clumsy, muck), and words beginning with sl also tend to have a negative connotation (slouch, slut, slime, sloven; Duduciuc, 2015, p. 113).

Canonical Composition

Citizens are more likely to vote for political candidates whose names are easy to pronounce (e.g., Mr. Smith vs. Mr. Colquhoun) because they misattribute this sensation: *Hmm, this name feels right. This candidate must be superior because [insert reason].*

In a follow-up study, researchers gave detailed information about candidates (e.g., previous career, outline of policies). Even with more information—*rational* information—people still voted in accordance with the name (Laham, Koval, & Alter, 2012). Guess I should thank my dad for my Polish name of Kolenda (and I'll send my apologies to the Hungarian psychologist, Mihaly Csikszentmihalyi).

But, perhaps an odd question: What makes a word composition more fluent? Consider two fictional words: BULEKA and KULEBA. Those names are similar, yet one is more fluent. Can you tell which one? Or why it's fluent?

In this section, you'll learn why most people prefer BULEKA. The answer involves your digestive system.

You experience a consistent bodily sensation: You swallow favorable objects, and you spit up unfavorable objects. Certain words can "feel good" because they mimic this sensation.

Figure 10D shows two columns of similar names, yet the names in Column 1 are more fluent. In those names, you pronounce the initial letters with the front of your mouth (e.g., B, M, P), and you pronounce the ending letters with the back of your mouth (e.g., K, R, G). Those compositions move *into* your mouth, so the experience feels like swallowing. The reverse happens in Column 2 (those compositions move out of your mouth).

In a collection of studies, people consistently preferred inward compositions across different languages (Topolinski et al., 2014).

Creating a brand name? Strive to create a brand name that moves from left to right across the spectrum in Figure 10E (Bakhtiari, 2015).

1	**2**
BULEKA	KULEBA
BALUGOR	RAGULOB
MESUKIRO	REKUSIMO
PATUGI	GATUPI
BATIKERO	RAKITEBO
PODAKERI	ROKADEPI

Figure 10D

				T		
				D		
				S		K
P				Z	CH	G
B				N	J	H
M	F			L	SH	NG
W	V	TH		R	Y	W

FRONT ⟵⟶ **BACK**

MENIKA outperforms **KENIMA**
FRONT ⟶ BACK   BACK ⟶ FRONT

Figure 10E

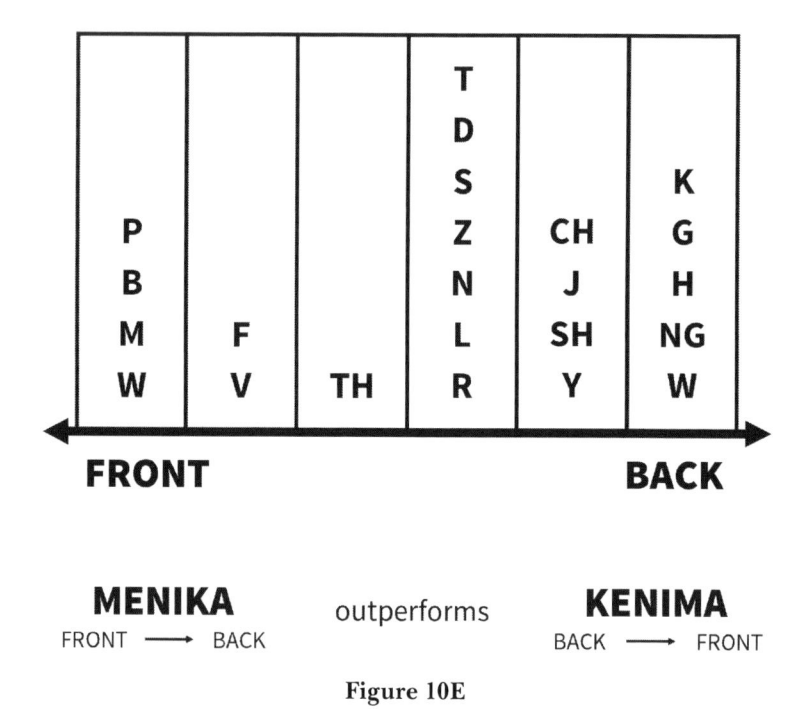

Canonical Stress

You also stress particular sounds within a word. Compare these sentences:

- ▶ The corlax cured the cow.
- ▶ The drugs corlax the cow.

Corlax is a fictional word. Yet, like most people, you probably stressed the word differently in both sentences. In the first sentence, you probably stressed the first syllable (CORlax); in the second sentence, you probably stressed the second syllable (corLAX; Kelly & Bock, 1988).

But why? Corlax is a fictional word, so nobody taught you the stress. The culprit is the part of speech.

For most words with two syllables, you find a pattern: Nouns stress the first syllable, while verbs stress the second syllable. In fact, changing the stress can change the part of speech: PERmit is a noun, while perMIT is a verb. Other examples: *record, compound, present, convert.*

We apply this stress for rhythmic alternation—that is, we alternate between syllables with and without stress. Since nouns often follow words *without* stress (e.g., "the" corlax), you stress the beginning of nouns to alternate the stress. The opposite happens with verbs, which often follow words *with* stress (e.g., the "drugs" corlax). You remove the beginning stress of verbs to maintain the alternating rhythm.

Other Primitives

You insert sensory traits into abstract concepts. Well, SOUND is an abstract concept—you can't see it, touch it, or feel it. Thus, you insert past primitives into this mental imagery.

Size

Sounds are intangible, but you conceptualize them with a spatial size. Louder sounds seem larger in magnitude:

- ▶ While viewing a rising sequence of numbers, people perceived a sound to get louder (Alards-Tomalin, Walker, Nepon, & Leboe-McGowan, 2017).
- ▶ People were quicker to respond while hearing small numbers spoken with a soft voice (Hartmann & Mast 2017).
- ▶ After hearing loud tones, people gave larger numbers (Heinemann, Pfister, & Janczyk, 2013).

You also associate SIZE with PITCH. Hit any object, and you'll hear a difference: Large objects produce low-pitched sounds, while small objects produce high-pitched sounds.

You just learned that LARGE is LOW PITCH, and you previously learned that LARGE is POWER. Simple math can derive a new connection: LOW PITCH is POWER.

You see this connection in animals: Dogs *growl* when they are aggressive, but they *yelp* when they are submissive.

Even humans succumb to this behavior: Whenever you summon help with a question, you typically raise the intonation of your voice (Ohala, Hinton, & Nichols, 1997). Also, in a mock dating competition, men spoke with a higher-pitched voice if they felt less dominant (Puts, Gaulin, & Verdolini, 2006).

Finally, which word conveys a large table: mil or mal?

Both words are fictional, yet most people choose mal because it simply "feels right" (Sapir, 1929). Some researchers blame this finding on mouth size (e.g., the word "mal" enlarges the mouth), but you can also blame the PITCH of each word: Back vowels (e.g., mal) are lower in pitch, so these objects seem larger; front vowels (e.g., mil) are higher in pitch, so these objects seem smaller (see Whalen & Levitt, 1995).

Object

Suppose that you hear a musical tone that is pulsing and disjointed. Which visual line resembles that tone: dotted or straight?

Dotted line, right? That question is so obvious that babies answer correctly (Wagner, Winner, Cicchetti, & Gardner, 1981).

But is it obvious? Sight and sound are two separate modalities, yet as you can see (or hear?), they share meaning. You derive this meaning because of the OBJECT primitive: You group the sensory world into discrete objects, and you extend this tendency into the domain of sound.

Location

Higher pitch is ... well ... higher. While listening to a high-pitched sound, people were quicker to press a button that was located in a high position (Rusconi et al., 2006).

Since you also associate UP and GOOD, you naturally connect GOOD and HIGH PITCH. Sure enough, after viewing a positive word, such as *kiss*, ambiguous sounds seemed higher in pitch (Weger, Meier, Robinson, & Inhoff, 2007).

Distance

What happens when you move toward a sound? It gets louder, right? You inject this DISTANCE into SOUND. Researchers asked people to guess the location of a caller on the phone. People believed that the caller was geographically closer when the volume was louder (Zhang, Lakens, & IJsselsteijn, 2015).

Perhaps you should avoid first dates in a loud atmosphere. Your shouting will reinforce a farther physical distance, which will reinforce a farther emotional distance.

Motion

Imagine that somebody is increasing the volume of a sound. You conceptualize this adjustment with sensory motion, so you simulate the momentum. Researchers confirmed this finding: A sound that was increasing in volume seemed louder at the end because of a momentum effect (Neuhoff, 2001).

Shape

In Figure 10F, I'm performing a *sheeb* task. So, take a guess ... what am I doing?

Figure 10F

This chapter opened with the bouba-kiki effect, which has been a mystery for decades. Researchers have described two explanations:

- ▶ **Oral Symbolism.** Bouba seems round because we pronounce that word with a round mouth. Kiki seems sharp because we pronounce that word with an angular mouth.
- ▶ **Perceptual Symbolism.** Sounds have meaning because of the written format of letters: The "b" in bouba is round, while the "k" in kiki is sharp (Lockwood & Dingemanse, 2015)

Despite those explanations, evidence is still conflicting. Some studies argue that the vowels are more meaningful (Spector & Maurer, 2013), whereas other studies argue that the consonants are more meaningful (Fort, Martin, & Peperkamp, 2015). Why the inconsistency? Perhaps we can blame PITCH.

High-pitched sounds are called SHARP tones because they produce a sensation of intensity that is similar to physical sharpness.

You find this intensity in *kiki* and *bouba*. The front vowels in kiki produce a high-pitched sound that you associate with sharpness; the back vowels in bouba produce a low-pitched sound that you associate with roundness. In a separate study on brand names, people preferred front vowels (e.g., brimley) for a knife, yet they preferred back vowels (e.g., bromley) for a hammer (Lowrey & Shrum, 2007).

Front vowels can also "sharpen" your focus. Remember the sheeb task? Researchers asked people to describe a similar image: A *sheeb* task elicited concrete behaviors (e.g., making a list), whereas a *shoob* task elicited broad behaviors (e.g., getting organized; Maglio, Rabaglia, Feder, Krehm, & Trope, 2014).

Okay, I'm done throwing connections at you. The next chapter will discuss another primitive concept that shapes your perception.

Sound good?

Summary

You attach meaning to sounds. You prefer sounds that:

- ▶ Move into your mouth
- ▶ Feel easy to pronounce
- ▶ Match prototypical qualities

SOUND is also intangible, so you conceptualize this imagery with sensory traits. Louder sounds seem larger in size, while high-pitched sounds seem physically sharp.

Other examples:

- ▶ **Canonical Speed.** I listen to a lot of YouTube videos at 2x the speed. Perhaps these videos are establishing a canonical pace of speaking, which could explain my tendency to talk fast in real life.

- ▶ **Canonical Accents.** You associate accents with ethnicities, so you expect to hear certain voices from people. Unexpected accents violate these expectations. In one study, participants exhibited a cardiovascular threat response while encountering an Asian person who spoke with an accent from the southern US (Mendes, Blascovich, Hunter, Lickel, & Jost, 2007).
- ▶ **Names of Politicians.** Researchers ranked the names of presidential contenders in all US elections from 1824. Candidates with the more pleasant name won in 35 of the 42 elections (Smith, 1998). Similar outcomes occurred with local elections in Washington, as well as Senate and House elections (see also Lowrey & Shrum, 2007).
- ▶ **Sounds and Physiology.** You are more likely to help a person named Kelly, rather than Kelsa, because her name ends in a hard "e" sound, which extracts a smile (Kniffin & Shimizu, 2016). *Hmm, do I want to help Kelly? Well, I feel a vague sense of smiling. So, yes.*

11

Physiology

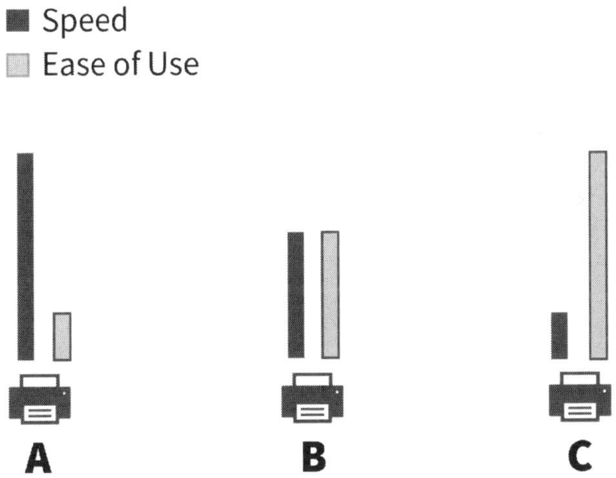

■ Speed
▢ Ease of Use

A **B** **C**

Figure 11A

STAND UP, and lift one foot into the air. Now, look at the printers in Figure 11A. Which printer would you choose?

You can sit down. In this chapter, I'll explain why you were biased to choose Printer B.

Weight

You are reading this book in a particular format: paperback, hardcover, or electronic. Each format has a particular weight, and this heaviness could influence your perception of the content.

Density

Heavy objects are typically larger and denser, so you establish a connection between WEIGHT and DENSITY. Heavy containers of food seem denser, as if more food is packed inside them (Piqueras-Fiszman & Spence, 2012).

Some companies distinguish their credit cards by making them heavier, creating a problem that I call the *heavy card effect*. Suppose that you are buying a laptop, contemplating whether to buy the extra insurance. While holding a heavy card, your bank account will literally seem bigger. And, with more funds (seemingly) available, you feel less pain with the purchase.

Currently, the US has accumulated more than $1 trillion in credit card debt (Comoreanu, 2018). If the heavy card effect is responsible for 1/10,000th of that amount, then we have generated $100 million in debt from the mere weight of a credit card. Perhaps we need a policy to control these weights.

Importance

You also associate WEIGHT with IMPORTANCE. Managers evaluate job candidates more favorably while reading their applications on a heavy clipboard (Ackerman, Nocera, & Bargh, 2010). Managers place more "weight" on those candidates because their applications seem densely packed with more credentials.

Some researchers had trouble replicating this effect, and they argued that this effect doesn't exist (Rabelo, Keller, Pilati, & Wicherts, 2015). But I have a few ideas why those studies didn't replicate.

First, you need a specific hypothesis for the importance; otherwise, a heavy stimulus will merely feel heavy. For example, researchers hid a weight inside the book *The Catcher in the Rye*, but the weight only strengthened the perception of importance for people who already read it. They needed enough knowledge about the book to generate a possible explanation: *Hmm, I feel a heavy sensation. Perhaps this book is more important because [insert reason]*.

Second, previous studies might have failed to replicate this finding because "importance" is misleading. Heavier objects aren't necessarily important . . . they're *impactful*. These concepts are distinct, as you'll see next.

Causal Potential

In the sensory world, heavier objects generate bigger effects.

> Being hit by a heavy object generally has more profound consequences than being hit by a light object, and the energetic costs of moving a heavy object are higher than those of moving a light object (Jostmann et al., 2009, p. 1169).

You insert this principle—*causal potential*—into WEIGHT. For example, medicine seems more effective when it's heavier (Kaspar, 2013a). It doesn't necessarily seem more important; it just seems more likely to generate an effect (e.g., getting better).

Similarly, water seems less favorable in a flimsy cup (Krishna & Morrin, 2007). Perhaps a flimsy cup generates a weaker effect on thirst.

Balance

After babies discover WEIGHT, they discover BALANCE: Any weight on the left side needs to be compensated by weight on the right side. This experience fueled abstract concepts, like stability and ambivalence.

Stability

You molded sensory stability into abstract stability. While sitting in a wobbly chair, people estimated that well-known couples (e.g., Barack and Michelle Obama) would break up sooner (Kille, Forest, & Wood, 2013). In another study, while standing on one foot, people believed that their own relationships would end sooner (Forest, Kille, Wood, & Stehouwer, 2015).

Sensations of physical instability—high heels, escalators, subways—can activate a desire for abstract stability. People who sat in a wobbly chair preferred romantic partners with stable traits (e.g., trustworthiness, reliability), compared to unstable traits (e.g., spontaneity, adventurousness; Kille et al., 2013).

Remember the printers from earlier? During that choice, you felt physically unstable. All else equal, you were more likely to choose Printer B because it was a compromise of two traits. It possessed the "balance" that you were seeking (Larson & Billeter, 2013).

Ambivalence

Ambivalent attitudes are positive *and* negative (*ambi* means both, while *valence* means positive or negative). Fast food tastes good, yet it's unhealthy. Reading is useful, yet it can be boring (except this book, of course).

Language is ripe with metaphors that describe ambivalence in terms of BALANCE.

> You waver BACK AND FORTH if you don't TAKE A SIDE. You might even reference your hands: ON ONE HAND, you expect this statement to convey a valence; ON THE OTHER HAND, you expect this new statement to convey the opposite valence. Meanwhile, you remain LEVEL-HEADED, EVEN-HANDED, and EVEN-TEMPERED.

BALANCE lies within ambivalence. In a clever study, participants read an article about eliminating the minimum wage: One version described positive and negative arguments, while another version described only positive arguments. While reading the ambivalent article, people were more likely to shift their weight from side-to-side. The reverse happened, too: While moving from side to side, people indicated more mixed feelings about a topic (Schneider et al., 2013).

Temperature

Pop quiz. You want to persuade somebody to perform a favor. What should you give this person?

- Cup of coffee
- Purple marker
- $1

Trick question. Every item should work because they all trigger a need to reciprocate (Burger, Sanchez, Imberi, & Grande, 2009).

But for our purposes, coffee is the real winner. People are more selfless when they hold hot coffee (vs. iced coffee) because the WARMTH activates AFFECTION (Williams & Bargh, 2008).

Temperature has other metaphors, too. Depending on the context, WAMRTH can become ANGER:

> Let's not get into a HEATED debate. You become HOT-HEADED or HOT-TEMPERED when your temper FLARES.

People were quicker to read the word "furious" in a hot font (e.g., fire at the top) compared to a cold font (e.g., ice at the top). In another study, an ambiguous face seemed angrier on a campfire background compared to an icicle background (Wilkowski et al., 2009).

It might seem odd that WARMTH is both ANGER and AFFECTION. Then again, maybe that's why anger can turn into sexual passion (see Borreli, 2015).

Smell

Temperature is also connected to smell: Cinnamon feels warm, while spearmint feels cold (Krishna, Elder, & Caldara, 2010).

But why do we like or dislike scents? And why do we disagree? Call me crazy, but I actually enjoy the smell of skunk.

Although a few smells, like ammonia, are always unpleasant because they stimulate the trigeminal nerve, most smell preferences are dictated by culture and expectations. Participants disliked a scent when it was labeled vomit, but other people enjoyed that same scent when it was labeled parmesan cheese (Herz & von Clef, 2001).

You also derive smell preferences from your first exposure to a scent:

> ...we acquire the emotional meaning of odors through experience, but first experiences are pivotal. This is why childhood, a time replete with first experiences, is such a training ground for odor learning. The first associations made to an odor are difficult to undo (Herz, 2001, p. 7).

Smell is strongly connected to the amygdala and hippocampus, which involve emotion and memory. When you smell a new scent, you inject your current emotions into this odor. Later, exposure to that scent activates those memories and emotions. In a lab study, people started preferring a scent after smelling this aroma in a positive context (Herz, Beland, & Hellerstein, 2004).

If you share my fondness for skunk smell, then—like me—perhaps your first exposure to this odor contained positive emotions (e.g., playing outside).

Canonical Smell

Surprise... you also have *canonical smells*. Certain objects—flowers, oranges, wood—possess a "typical" smell, and you prefer smells that match these aromas. For example:

- ▶ **Temperature.** Cinnamon enhanced a warm gel pack, while spearmint enhanced a cold gel pack (Krishna et al., 2010).
- ▶ **Shape.** People preferred angular shapes while smelling lemon and pepper, yet they preferred round shapes while smelling raspberry and vanilla (Hanson-Vaux, Crisinel, & Spence, 2012). The intensity of lemon and pepper matched the intensity of angular shapes (and vice versa).
- ▶ **Texture.** People preferred masculine scents on rough paper, yet they preferred feminine scents on smooth paper (Krishna et al., 2010). I'll expand on gender later.

Suspicion

Today is your lucky day, my friend. I'm going to mail you $10,000.

What's that? You don't believe me? Alright, you caught me. I'm lying. But I wanted to illustrate that—right now—you can more easily detect the smell of fish.

First, you should realize that it can be difficult to identify scents. You might recognize a scent, but you often struggle to label the identity.

> When a person says "I smell something," it usually means she or he detects an odor but cannot identify it with certainty . . . Just like suspicion, it takes time to figure out what a smell is, and people may or may not find out in the end (Lee & Schwarz, 2012, p. 747).

Many languages describe suspicion in terms of smell because of these similarities. In English, we say that something "smells fishy."

Therefore, we associate suspicion with fish smells. Researchers sprayed various scents in a school hallway: water, fish oil, or fart spray (and yes, fart spray was the technical term). Then, they recruited people to play an economic game: Researchers gave them $5, and they could share however much—if anything—with another student. That money would quadruple, and the opposing student could decide however much—if anything—they would share in return. In this setup, people give less money if they distrust the opposing student. And, indeed, they gave less money if they smelled fish (but not water or fart spray; Lee & Schwarz, 2012).

The reverse happened, too: Exposing participants to suspicious words (e.g., distrust, shady, uncertain) influenced them to complete word fragments (e.g., FI_ _ING, F_N, TU_ _) with fishy words (e.g., FISHING, FIN, TUNA).

Perhaps most interesting, suspicion helped participants detect a fish smell. They smelled five scents—apple, onion, caramel, orange, fish—while trying to identify each scent. In some trials, researchers acted suspiciously by saying: "Obviously, it's a very simple task and you know, there's... there's nothing we're trying to hide here." After that suspicious behavior, participants could more easily identify one scent: fish.

Touch

See the person in Figure 11B? Is he a professor of physics or history? Keep your guess in mind.

You gain information from touching objects (see Figure 11C; Lederman & Klatzky, 1987). You also distinguish particular traits of touch, like texture, which you associate with different ideas.

For example, you associate TEXTURE with DIFFICULTY:

> You use COARSE language on a ROUGH day—but if you're having a good day, then it's SMOOTH sailing.

Figure 11B

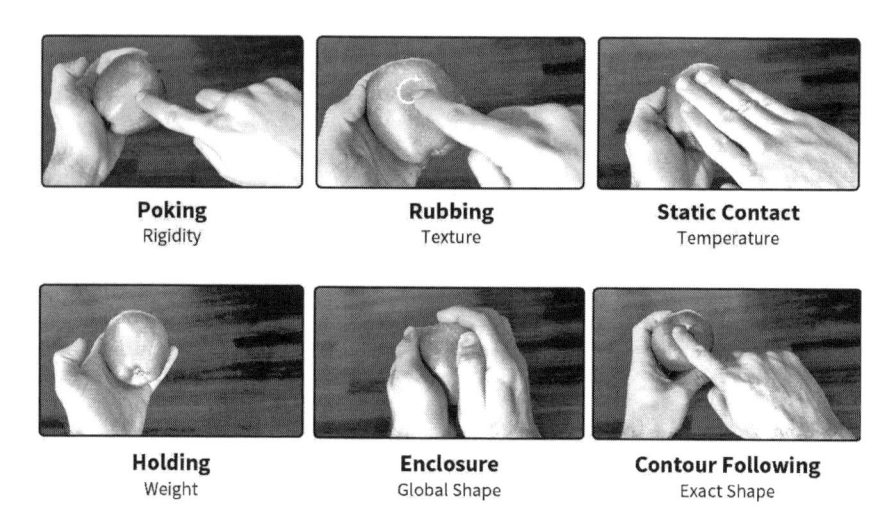

Poking Rigidity	**Rubbing** Texture	**Static Contact** Temperature
Holding Weight	**Enclosure** Global Shape	**Contour Following** Exact Shape

Figure 11C. Haptic cues that provide information.

Same with RIGIDITY. Participants who sat in a hard chair made fewer concessions during a negotiation because the sensory rigidity activated an abstract rigidity (Ackerman et al., 2010).

These metaphors also influence your thinking style. People solved more anagrams while sitting in a soft chair because they possessed a flexible mindset; however, people solved more memory puzzles while sitting in a hard chair because they possessed a rigid mindset (Xie, Lu, Wang, & Cai, 2016).

Remember your guess about the professor? Rigidity also appears in academic fields: HARD sciences contain rigid principles, while SOFT sciences are flexible. When people squeezed a rigid ball, they guessed that somebody was a professor of physics (vs. history; Slepian, Rule, & Ambady, 2012).

What chair are you sitting in now? Your sensory experience might have tainted your perception of that picture.

Summary

You feel physiological sensations, and you insert these traits into other ideas.

- ▶ **Weight.** Heavy objects seem bigger, denser, and more impactful.
- ▶ **Balance.** Balance created abstract stability and ambivalence.
- ▶ **Temperature.** Warmth activates affection, yet also anger.
- ▶ **Smell.** You prefer scents that contain positive memories.
- ▶ **Texture.** Sensory roughness activates abstract roughness.
- ▶ **Rigidity.** Sensory rigidity activates abstract rigidity.

You can see all of those traits in this book. Sensory traits of this book are influencing your perception of the content:

- ▶ **Physical vs. Electronic.** Reading the paperback? During the first half, the right side of the book will weigh more because it comprises the majority of pages. Once you cross the halfway point, however, the left side will weigh more. While evaluating the content, you might place more weight on whichever side physically weighs more.
- ▶ **Hardcover vs. Softcover.** All of these ideas might seem rigid and infallible in a hardcover book, yet they might seem flexible and adaptable in a softcover book.
- ▶ **Smooth Cover vs. Rough Cover.** Smooth surfaces facilitate motion. Perhaps you assume that you can move through this book more easily with a smooth cover. A rough cover would imply tougher information inside.

12

Emotions

SOME DOCTORS TREAT depression through Botox injections (Magid et al., 2015). Ridiculous, right? Or . . . maybe not. Can you spot the explanation?

As babies grow older, they start feeling a wide mixture of emotions. This chapter will explain what emotions are, why we have them, and how they influence your perception of the world.

Purpose of Emotion

Why do humans feel emotions? Emotions must have helped our ancestors in evolution, but how? I'll argue that emotions were adaptive in three ways.

1. Bodily Adaptations

Emotions adjust your body in useful ways. Women feel disgusted during the first trimester of pregnancy because their immune system is weaker—so it protects them (and their babies) from foodborne illness (Fessler, Eng, & Navarrete, 2005).

Disgust also crinkles your nose to block sensory input:

> ... facial expressions are not arbitrary configurations for social communication, but rather, expressions may have originated in altering the sensory interface with the physical world (Susskind et al., 2008, p. 843).

2. Behavioral Adaptations

Emotions were helpful with social behaviors, too. Anger helped our ancestors improve their welfare in society. If people felt entitled to more welfare, they showed anger to enforce this change.

Who felt entitled? People with greater bargaining power:

- ▶ People who could inflict more costs (e.g., powerful brutes)
- ▶ People who could give more benefits (e.g., attractive mates)

Those people had greater leverage, so they displayed anger—even in fair transactions—because they expected better treatment. Unfair? Perhaps. But you can blame Darwin.

More importantly, humans still possess those adaptations. Even today, certain people are more prone to anger: stronger men and attractive women (Sell et al., 2009). In fact, politicians are more likely to use military force when they're physically stronger (Sell et al., 2009). Based on their personal simulations of the world, they falsely believe that other nations will be less likely to retaliate. Bet you feel safer now, huh?

3. Perceptual Adaptations

Emotions help you perceive stimuli in the most adaptive way possible, a principle that I call *adaptive perception*.

For example, studies show that positive stimuli seem larger (Veltkamp et al., 2008), while other studies show that negative stimuli seem larger (van Ulzen, Semin, Oudejans, & Beek, 2008). Why the contradiction?

You are perceiving both stimuli in the most helpful way: Positive stimuli seem larger, closer, and more desirable so that you're motivated to acquire them, whereas negative stimuli seem larger, closer, and more dangerous so that you're motivated to avoid them. I call it adaptive perception because of the dual meaning: You *adapt* your perception toward your current needs, and this mechanism was *adaptive* in evolution.

Emotions distort your perception so that you become motivated to act. Thirsty people perceive water bottles to be closer and larger so that they become motivated to acquire them (Veltkamp, Aarts, & Custers, 2008). In another study, rich and poor children shared the same perception of size for round cardboard discs, yet poor children perceived coins—a desirable stimulus—to be physically larger (Bruner & Goodman, 1947).

Likewise, threatening stimuli seem worse. While standing at the top of a hill, the decline seems steeper on a skateboard (vs. wooden box) because this distortion discourages a risky behavior (Stefanucci, Proffitt, Clore, & Parekh, 2008).

Similar effects occur with memory. In one study, people listened to the irritating sound of a vacuum. If they needed to listen to that sound again, the past memory seemed worse to discourage this event from happening again (Galak & Meyvis, 2011).

Finally, you also adapt your attention toward stimuli. For example, our female ancestors needed motivation to reproduce while ovulating, a time period in which conception was more probable. Lo and behold, ovulating females are quicker to notice male faces (Macrae, Alnwick, Milne, & Schloerscheidt, 2002). They also prefer masculine features—big jaws, prominent cheekbones, and even body odor (Penton-Voak, & Perrett, 2000; Grammer, 1993).

There's a common saying: You see what you want to see. But that's inaccurate. You see what you NEED to see.

- ▶ **Adaptive Perception.** You see what you need to see.
- ▶ **Adaptive Memory.** You recall what you need to recall.
- ▶ **Adaptive Attention.** You notice what you need to notice.

In sum, humans feel emotions because these responses helped our ancestors survive.

What is Emotion?

Until now, I've been treating emotion—anger, fear, sadness—as discrete categories. But there's a problem:

> Even after a century of effort, scientific research has not revealed a consistent, physical fingerprint for even a single emotion (Barrett, 2017, p. xii).

Wait, what? Emotions aren't real? Well, not in the conventional sense. In fact, some languages don't even have a word for "emotion" (Russell, 1991). English has the word "emotion" because we developed that terminology before we understood this concept. Whoops. Some researchers are trying to change the terminology (see Russell, 2003). But meh, it's probably too late.

So then, what is emot—err, those "things" we feel? Think of emotion like color. Color is a spectrum with infinite variations. We can't label every color, so we categorized the spectrum into overall hues (e.g., red, green, blue). Ultimately, these categories are entirely subjective. English has a single hue for blue, while the Russian language categorizes this hue as two different colors: lighter blues (*goluboy*) and darker blues (*siniy*; Winawer et al., 2007).

Emotion works the same way. You've probably heard that humans feel six universal emotions: fear, disgust, anger, happiness, sadness, and surprise (Ekman & Friesen, 1971). But those categories are

Emotions 141

subjective and arbitrary. Just like some cultures categorize the color spectrum differently, cultures can categorize the emotion spectrum differently, which shatters the idea of universality.

> If English language categories regarding emotion are not universal, then we have no guarantee that emotion, anger, fear, and so on are labels for universal, biologically fixed categories of nature. Rather, they are hypotheses formulated by our linguistic ancestors (Russell, 1991, p. 444).

Researchers have tried to replicate the studies on universality, but their findings don't support a compartmentalized notion of emotion (even though "notion of emotion" sounds really cool). In those replications, remote populations can't identify discrete emotions from supposedly "universal" facial expressions (Gendron, Roberson, van der Vyver, & Barrett, 2014a).

Our mistaken framework of emotion sent researchers down a long and windy hole that we're only beginning to escape.

So then, if emotion isn't universal, which—if any—emotions are real? And what exactly *are* those feelings?

Your emotions arise from *core affect*.

Core Affect

You are feeling core affect right now. It's a constant physiological state with two components:

- **Valence:** Good or bad.
- **Arousal:** Calming or energizing.

You can categorize *all* emotions with these two dimensions (see Figure 12A).

HIGH AROUSAL

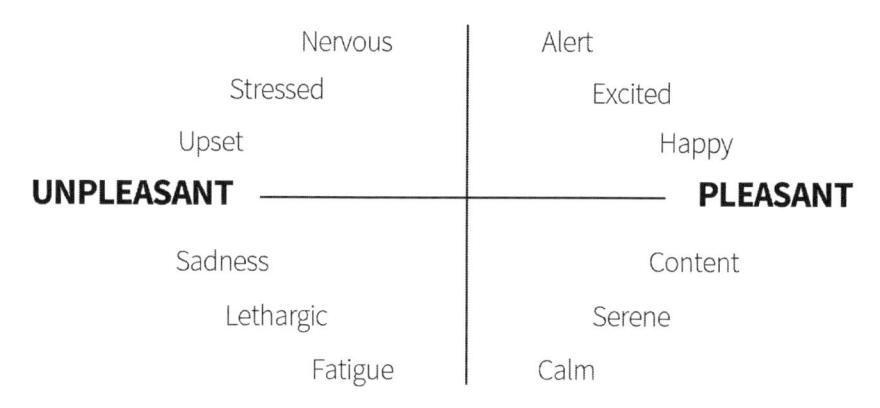

Nervous	Alert
Stressed	Excited
Upset	Happy

UNPLEASANT ——————————————— **PLEASANT**

Sadness	Content
Lethargic	Serene
Fatigue	Calm

LOW AROUSAL

Figure 12A. Most emotions can be captured with two dimensions.

In the old paradigm, a neutral state was the absence of emotion. But you are *always* feeling emotion. Even right now. If no feelings are particularly strong, you possess a neutral state of valence and arousal.

That distinction is important because it changes the implications of emotion. In the old paradigm, fear activated a distinct fear module. Yet, in reality, there is no fear module. Fear activates different regions depending on the source of fear, such as physical danger (e.g., midcingulate cortex, bilateral posterior insula, parahippocampal cortex) or social evaluations (e.g., ventromedial prefrontal cortex, left inferior frontal gyrus, posterior occipital cortex; Wilson-Mendenhall, Barrett, Simmons, & Barsalou, 2011).

Researchers had been studying emotions as if they were discrete biological underpinnings, yet emotions don't exist in these discrete forms. Humans created these emotions through language.

If these emotions don't exist, why do they feel so real? The answer involves scaffolding.

Scaffolding of Emotion

Sensory concepts fueled your knowledge. You took a vast array of sensory traits, and you extended these concepts into many new ideas.

Emotions followed a similar trajectory. Babies enter this world with two reflexive emotions—contentment and distress—and they expanded these feelings into similar (yet more complex) emotions, such as happiness and sadness. And they kept expanding these emotions into finer versions, such as anger, fear, and awe.

Language was the culprit behind these extensions. You feel emotions *because* of the terms we created.

Here's an example from a different domain. In Chapter 8, I described the Korean term—kkita—which depicts the tightness of containment. English speakers don't learn this word, but we can still perceive tight containers, right? Surely we don't need a term? Well, yes and no. Linguistic terms facilitate this process: Toddlers are quicker to grasp the concept of kkita when they are taught a novel word for this idea (Casasola, Wilbourn, & Yang, 2006).

Emotion works the same way. Specific terms—anger, fear, awe—help you notice these feelings when they arise.

Let's try it out.

You probably feel good after learning something new. You encounter this sensation, but you rarely notice or acknowledge that feeling when it arises because you don't possess a term for it. We can change that:

> **Brainified**—When you feel good or satisfied after learning something new.

Voila. Now that you possess a term, you might notice instances of brainification more easily.

To recap, humans aren't endowed with universal emotions like we previously believed. Humans feel core affect, a constant fluctuation of valence and arousal, and we label these fluctuations based on terms that we derive from culture and language.

And that, my friend, is emotion. Are you tired? I'm tired. But now that we covered the foundation, let's see the implications.

Simulating Emotion

You activate past experiences whenever you simulate concepts. If you think of a bird, you activate a chirping sound (Mulatti et al., 2014).

Well, emotions are conceptual in nature. Whenever you conceptualize an emotion (e.g., feel the emotion, see the emotion in somebody else), you activate past experiences with this concept.

Consider an emotion, like happiness. When you feel happy, what do you do? You often smile, right? Therefore, your concept of happiness contains the physiological trait of smiling. Much like you simulate a chirping sound while thinking of a bird, you simulate a smile whenever you conceptualize happiness.

This insight can solve the mystery at the beginning of the chapter. Doctors are treating depression with Botox because these injections prevent people from frowning. Unable to frown, people struggle to simulate the emotion of sadness (Magid et al., 2015). In other studies, people with Botox were slower to understand sentences about anger and sadness, but they were unaffected with sentences about happiness (Havas, Glenberg, Gutowski, Lucarelli, & Davidson, 2010).

Heavy Emotions

If you look closely at language, you'll notice that we describe emotion with physical weight:

You feel UPLIFTED from a LIGHTHEARTED comedy, and you feel WEIGHED DOWN by a HEAVY drama.

Which emotions are heavy? I suspect that heavy emotions have low arousal and negative valence.

Negativity and low arousal hinder your ability to interact with the world. Hills seem steeper for people who are listening to sad music because they have trouble simulating the climb (Schnall et al., 2010). Sad music makes you feel "weighed down" because you have trouble simulating motor actions, as if you are being physically weighed down.

That's why keeping a secret can feel like a "heavy" burden:

> ...when preoccupied by a secret, one is devoting personal resources toward that secret. This increased preoccupation with the secret might suggest to the secret holder that increased effort is needed to keep the secret (and thus less effort is available for other pursuits; Slepian, Camp, & Masicampo, 2015, p. 3)

The reverse happens with forgiveness, an action that ends preoccupation (and lifts a psychological weight). After people forgave somebody, they jumped higher in a fitness test (Zheng et al., 2015).

Emotion and Attention

You might have encountered the wildly popular book *The Secret* that makes a bold claim: If you wish something to happen, then it *will* happen. If you're a fan of that book, I have good news and bad news.

First the bad: The explanation is bogus. Whenever self-help gurus describe quantum physics, you should usually run away.

But here's the good news: The underlying mechanism *does* work (to some extent). But the real "secret" doesn't involve weird energy. It involves your emotions.

Holding a spherical object will orient your attention toward other spherical objects, such as an orange. Emotion works in a similar way: After you see positive pictures, you can process other positive stimuli more easily because your brain is attuned to this positive valence (Spruyt, Hermans, Houwer, & Eelen, 2002).

If you focus on positive emotions—and what you desire—you're more likely to notice positive stimuli in the world around you. You aren't changing the objective reality; you are merely changing your perception of reality.

Or, in some cases, you *could* change reality through a self-fulfilling prophecy. If you believe that you will achieve a high score on your next exam, you might follow behaviors (e.g., studying) to achieve that goal. So yes, your thinking *can* help you achieve what you want, but don't thank the magical voodoo that altered the universe.

We should promote positive thinking in society, but we shouldn't propagate a misleading explanation. Not only does *The Secret* denigrate science, but it also contains a dual message that negative events—rape, murder, bullying—are the faults of victims . . . if only they had wished more. We need to squash that philosophy.

Okay, my rant is done. Let's jazz things up with COLOR.

Summary

Across evolution, humans acquired a combination of valence and arousal. These feelings distorted our perception in helpful ways.

You feel *many* emotions because we categorized the emotion spectrum into discrete terms. Every emotion—anger, fear, awe—feels unique *because* of this linguistic foundation. These terms helped you notice these ideas whenever they occurred, and you inserted these ideas with your own experiences. Today, every time that you feel an emotion, you activate this cluster of experience.

Other examples:

- **Adaptive Expectations.** You predict what you *need* to predict. In one study, people competed against a professor on a motor skill, and they estimated their potential happiness if they beat the professor. When people estimated before the game, they predicted higher levels of happiness. When people estimated after the game, yet before learning the results, they predicted lower levels of happiness. They needed to predict stronger emotions before the game to extract better performance (Morewedge & Buechel, 2013).
- **Body States.** Emotions activate body states (e.g., happiness activates smiling). Yet sometimes body states can activate emotions. When researchers asked people to furrow their eyebrows, a body state associated with thinking hard, they rated celebrities to be less famous (Strack & Neumann, 2000). *Hmm, how famous is this person? Well, I'm furrowing my eyebrows, so I must be thinking hard to remember them.*
- **Product Descriptions.** Writing a product description? You should activate the target emotions that buyers are seeking. While buying a nonfiction book, potential buyers might ask: Will I learn something? And, if they learn something while reading the description, they conclude: *Hmm, will I learn something? I just learned something now. So, yes.* While buying a fiction book, they might ask: Will I *feel* something? Again, your description should elicit these feelings.

13

Color

| Out of Cereal | Fistful of Cake |

Figure 13A

YOU CAN SEE two pictures of me in Figure 13A. I slightly increased the brightness in one picture. Can you tell which one?

Got your choice?

Actually, I didn't change the brightness. I wanted to illustrate that you were more likely to choose the cake picture. We'll revisit the answer later.

You are engulfed in a sea of color. Surely research on color must exist, right? A lot of research, presumably?

You'd think so, but color psychology is a scarce topic. To put things in perspective, I searched various keywords on Google Scholar, and I tallied the number of studies that appeared (see Figure 13B).

Number of Studies

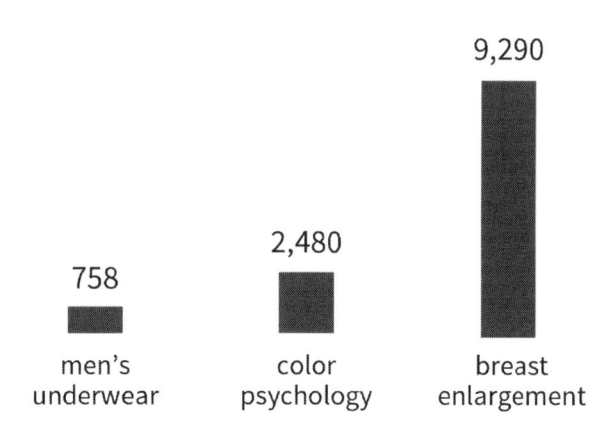

```
                                          9,290
                                          ████
                                          ████
                                          ████
                                          ████
                          2,480            ████
            758           ████             ████
             ▬            ████             ████
        men's            color            breast
      underwear        psychology      enlargement
```

Figure 13B. Color psychology is near the same academic rigor as men's underwear.

Luckily, the topic seems to be growing since I compiled that data in 2015, so hopefully the momentum continues.

What is Color?

Color has three components:

- ▶ **Hue.** Labels in the color spectrum (e.g., red, blue, green)
- ▶ **Lightness.** Colors can be dark or light
- ▶ **Saturation.** Colors can be vivid or faded

Ultimately, color hues—like emotion—were built from language. Humans categorized the color spectrum into arbitrary distinctions (e.g., red, green, blue), and we filled these terms with meaning. Today, seeing the color yellow produces the same neurological effect as reading the word "yellow" (Simmons et al., 2007).

Color Preferences

What's your favorite color? And why? Sure, it looks pretty . . . but *why* does it look pretty? You probably can't articulate that reason, so this section will explain what's happening inside your brain.

You can trace some color preferences to the needs of our ancestors. Female ancestors were foragers, so they needed to identify red fruit on green foliage. And, lo and behold, female color vision is sensitive to red colors on green backgrounds (Alexander, 2003).

However, evolution can't explain individual differences. Some people like purple, while other people like green. Why do these differences emerge?

Researchers uncovered the answer—it's called *ecological valence theory* (Palmer & Schloss, 2010). All of your color preferences have emerged from your past experiences with colors:

> The more enjoyment and positive affect an individual receives from experiences with objects of a given color, the more the person will tend to like that color (Palmer & Schloss, 2010, p. 8878).

If your favorite color is blue, then you've had the most positive experiences with blue. If your second favorite color is green, then you've had the second most positive experiences with green. And so on.

That claim might sound radical, so let's deconstruct the methodology behind that research. In the study, researchers asked participants to rate how much they preferred 32 colors, and they plotted their preferences on a graph. Most people gave the highest ratings to bluish colors, followed by reddish colors. The worst ratings were dark yellow and brown.

At this point, researchers created a variable—called Weighted

Affective Valence Estimate (WAVE)—to replicate that distribution of color preferences. Stick with me. First, they asked new participants to list objects associated with the 32 colors. For dark blue, somebody could write ocean. After listing those objects, people rated two things: (a) how much they liked each object, and (b) how closely the object matched the given color. From here, the researchers merged those responses into a singular variable—WAVE—and, by golly, it matched the original color preferences eerily well.

Using this WAVE variable, researchers could predict the color preferences of new people. If somebody really liked oceans—and if they indicated that oceans strongly resemble dark blue—then they strongly preferred dark blue.

Since then, this theory has been supported in lab studies. Researchers conditioned people to feel positive emotions with certain objects, and these participants started preferring the colors in those objects (Strauss, Schloss, & Palmer, 2013).

In sum, the entirety of your color preferences, from most favorite to least favorite, aligns with your past experience with each color.

As you'll see next, this mechanism also determines the semantic meanings of color.

Color Meanings

Many people assume that every color has a specific meaning: Blue is calm . . . yellow is cheerful . . . green is environmental.

But it's not that simple. Much like color preferences, color meanings are based on individual experience, which varies from person to person.

Nevertheless, humans live in the same world. Some colors are seen in similar contexts across the globe, so a handful of meanings are bound to be universal.

This section will explain those universal meanings.

Temperature

Across the world, all sources of heat (e.g., fire, sun) are RED, ORANGE, and YELLOW. Not surprisingly, we refer to these colors as warm. Cool colors are BLUE, GREEN, and PURPLE.

These metaphors influence your perception, too. Warm colors make objects seem physically warmer—for instance, coffee seems warmer in red cups (Guéguen & Jacob, 2014).

Whenever you feel physically cold, you might evaluate warm colors more favorably because they provide the warmth that you're seeking.

Aggression

You just learned that WARMTH is RED, and you previously learned that WARMTH is ANGER. Therefore, you bind RED and ANGER. Participants were quicker to read angry words in red fonts (Fetterman, Robinson, & Meier, 2012).

Using data from the 2004 Olympics, researchers analyzed matches in boxing, wrestling, and taekwondo. Despite a random assignment of uniform colors—red or blue—red competitors were more likely to win (Hill & Barton, 2005). Over a span of 55 years in England, football teams with red uniforms were more likely to be champions (Attrill, Gresty, Hill, & Barton, 2008).

Obviously, *something* is occurring—but which is it?

- Do players behave more dominantly in red?
- Do referees perceive more dominance in red?

Both.

First, players behave more dominantly. Participants exerted more force in a hand clasp if they were assigned a red ID number (Elliot & Aarts, 2011).

Second, red also *conveys* more dominance. Referees watched a video of a sparring match, and the researchers digitally swapped uniform colors. Both videos depicted the same match, yet referees awarded more points to whoever was wearing red (Hagemann, Strauss, & Leißing, 2008).

Attraction

Many animals are sexually attracted to red—male monkeys, for example, are attracted to female monkeys with red swelling in their butts because this color indicates a desire for sex (Dixson, 1983).

Humans inherited a similar mechanism. In an interesting study, women attended lab sessions for a couple months, and the researchers collected their saliva (while secretly noting their clothing). Turns out, these women were more likely to wear red during the fertile windows of their menstrual cycle (Eisenbruch, Simmons, & Roney, 2015). They were also more likely to wear red shirts if they expected to talk with attractive men (Elliot, Greitemeyer, & Pazda, 2013).

Women wear red to convey sexual interest, but men also believe that women in red are more interested in having sex. And these women are perceived to be more attractive (Pazda, Elliot, & Greitemeyer, 2012; Elliot & Niesta, 2008). Again, that mechanism was advantageous in evolution. If women displayed red to convey sexual interest, then men who noticed this gesture (and became attracted) would have been more likely to reproduce.

Despite the evolutionary foundation, however, perhaps a simpler explanation can be found in metaphors. You inserted a primitive idea of WARMTH into social warmth. Therefore, people who crave social warmth are unknowingly craving physical warmth. Red shirts will be more appealing because this color has the WARMTH that people are seeking.

Construal Level

You connected WARMTH and AFFECTION because you felt these concepts whenever you were held as a baby. But look closely. You can spot a third concept in this experience . . . PROXIMITY.

> . . . human bodies are warm, but one must be close to a body to feel its warmth . . . warmth implies spatial proximity to the heat source (Fay & Maner, 2012, p. 1369).

Therefore, you also connect WARMTH and PROXIMITY.

Now, you might recall that PROXIMITY triggers a low construal: If you are close to something, you focus on the details.

Warm colors, like red, should trigger this effect. Sure enough, when people evaluated a camera on a red background, they preferred detailed information (e.g., type of lens). On a blue background, they preferred broad information (e.g., potential uses; Mehta & Zhu, 2009).

Currently, researchers are perplexed why synesthesia patients associate the letter "A" with red (Simner et al., 2005). Perhaps we can blame RED and PROXIMITY: The letter "A" activates red because it is the first letter of the alphabet (and thus closest to people).

Valence

Light helps you see the world. It's no surprise that many cultures describe light in terms of knowledge:

> Let's ILLUMINATE this idea so that I don't KEEP YOU IN THE DARK. Do you SEE what I mean? Hopefully, by now, you SEE THE LIGHT.

Light also feels good because it reduces ambiguity. Eventually, you conclude: LIGHT is GOOD, while DARK is BAD.

You look on the BRIGHT SIDE when you have a RAY OF HOPE. You feel bad on GLOOMY days, but you remind yourself that it's always DARKEST BEFORE THE DAWN.

And these connections influence your perception. Researchers displayed two faces—smiling or frowning—and they asked people to choose the brighter image. Even though both images were identical in brightness, most people chose the smiling face (Song, Vonasch, Meier, & Bargh, 2012).

Remember the two images at the beginning of this chapter? You were more likely to perceive the positive image (the cake) to be brighter.

Darkness is especially connected to hopelessness. When people sit in a dark room, they indicate less hope for their future (Dong, Huang, & Zhong, 2015). And the reverse happens: Rooms seem darker while reflecting on a past memory of hopelessness.

But here's a question: Why hopelessness? How does this concept differ from a general bad feeling? Let's break it down.

If you feel hopeless, you can't see a bright future ahead—you are gazing across a distance of time, which is built with space. If you can't see a good outcome across a distance of time, then you can't see a good outcome across a distance of space. And thus, you perceive the environment to be darker. I suspect that a similar effect occurs with ambiguity: People who don't SEE what you mean—people who are metaphorically IN THE DARK—might perceive rooms to be darker.

We can expand this concept through color. Light is similar to white, and darkness is similar to black. Therefore, these colors will inherit the connotations of good and bad: Heaven is WHITE, while death is BLACK. Researchers confirmed this finding with font colors: Participants responded faster to positive words in a white font (yet negative words in a black font; Meier, Robinson, & Clore, 2004; Meier, Fetterman, & Robinson, 2015).

Unfortunately, this concept can also taint the perception of race. Researchers altered the skin tone of a biracial politician, and they asked people to choose the best photo. People chose a lighter skin tone when they supported his views, yet they chose a darker tone when they opposed his views (Caruso, Mead, & Balcetis, 2009). The same effect happened with pictures of Obama: Liberals chose lighter skin tones, while conservatives chose darker skin tones.

Morality

You also insert WHITE and BLACK into the moral domain. While reflecting on a memory in which you did something immoral, you perceive the room to be darker. You even prefer light-emitting objects, such as candles and lamps (Banerjee, Chatterjee, & Sinha, 2012).

Dark colors can also activate immoral behavior. Sports teams with black uniforms receive more penalties (Frank & Gilovich, 1988). It even occurs when teams switch uniforms: The Pittsburgh Penguins averaged 8 min per penalty with blue uniforms, but after switching to black uniforms, they started averaging 12 min per penalty.

Why does black trigger unethical behaviors? First, you should realize that we feel pressured to perform moral behaviors when other people are watching us. For example, customers are more likely to buy environmental products while shopping in public (Griskevicius, Tybur, & Van den Bergh, 2010).

Even if nobody is present, this effect can be activated with eyes. In a classic study, customers paid for coffee through an "honesty box" where nobody would notice whether or not they paid. Over several weeks, researchers alternated two images—flowers or eyes—and customers paid 3x more money in the presence of eyes (Bateson, Nettle, & Roberts, 2006).

Maybe this insight could explain why UP is MORAL. For example, people who went up an escalator were more likely to donate money

(Ścigała & Indurkhya, 2016). Perhaps upward locations feel more visible from various directions (and thus we need to act morally).

So, let's recap. You behave morally when other people are watching, a situation that could occur from a bright light. And, as you learned, a bright light is similar to the color white. Therefore, white activates moral behavior because you sense that other people will see this behavior. Black triggers immorality because it feels like nobody can see this behavior. Just like many people falsely assume that black cars hide dirt, you believe that darkness will hide your metaphorical impurities.

Background colors might be more influential than we thought. Some platforms (e.g., iPhones, Facebook) allow users to choose a white or dark theme. Perhaps these themes are influencing behavior. Perhaps users are more likely to post offensive comments in dark themes because they falsely believe that nobody will see them.

That mechanism could also shed light (see what I did there?) on patients with synesthesia. Researchers are perplexed why patients associate letters with colors, such as O with white, and X with black (Simner et al., 2005). Can you spot a reason? Circular shapes, like O, facilitate our vision—for example, we often circle something to emphasize it. Conversely, we use X's to cross things out. Perhaps O is WHITE (while X is BLACK) because of a deep metaphor with vision.

Color and Other Primitives

Finally, you bind COLOR with other primitives, such as WEIGHT and SOUND.

Weight

Did you notice that we describe color in terms of weight? Dark colors are heavy, while light colors are . . . well . . . light.

Color 159

Figure 13C

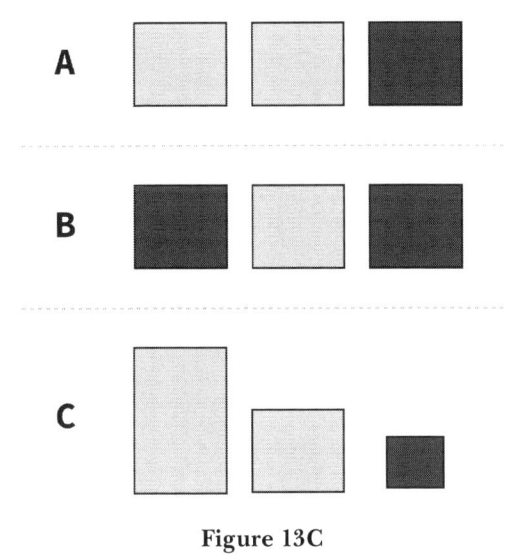

Look at Row A in Figure 13C. Doesn't it feel like the darker square is heavier, as if it's pulling the right side downward?

In order to stabilize the design, we need a counterbalance on the left, such as an equally dark square (Row B). But we don't need symmetry—we could also adjust the sizes (Row C).

Heavy stimuli sink downward. Perhaps retailers should position dark packages on bottom shelves (Sunaga, Park, & Spence, 2016). While buying coffee at Target, I noticed that light roast coffee was located on the bottom shelf, but this location is incongruent with lightness (see Figure 13D).

Size

You also connect COLOR with SIZE. Dark objects seem bigger, while bright objects seem smaller (Walker & Walker, 2012).

This finding is important because it provides the final piece of a puzzle that we haven't solved yet. Let's solve it now.

Figure 13D

Sound

I still haven't answered a key question from earlier in this book: Why do people associate brightness with sneezing?

Brightness and sneezing are separate concepts, yet they share a series of hidden connections underneath the surface. We can trace this lineage: Bright colors are visually lighter . . . visually light colors are physically lighter . . . physically lighter stimuli are smaller . . . and smaller stimuli generate a higher pitched sound.

We started with BRIGHT, and we traced this concept to HIGH PITCH. You associate brightness with the high pitch of a sneeze because of those hidden connections.

Summary

Humans categorized the color spectrum into specific hues (e.g., red, blue, green). Ultimately, these labels are subjective and arbitrary.

Working on a design project? Wondering which color to choose? Don't stress over this decision—most hues don't have a consistent meaning. Most associations of color will be found in broader categories of hues, such as warm or cool. Warm colors are more appropriate for:

- Physical warmth (e.g., coffee)
- Passionate emotions (e.g., affection, anger)
- Detailed information (e.g., product specifications)

Color has other categories, such as lightness or saturation—which are often more meaningful than hue. For example, darkness is associated with a negative valence and immorality.

Other examples:

- **Music and Brightness.** Shoppers prefer shelves with brighter décor while listening to high-pitched music (Hagtvedt & Brasel, 2016).
- **Book Covers.** Searching for a book that is light and readable? You will prefer a white cover with minimal text and graphics; this visual lightness conveys lightweight content. Want a book with dense information? You will prefer a dark cover because this visual heaviness conveys informational heaviness.
- **Driving.** Could the colors of cars influence driving behavior? Perhaps you are less likely to tailgate red cars because they seem closer to you. Or you might speed in black cars because the dark color hides your unethical driving (and you believe that cops are less likely to see you).
- **Red and Aggression.** People bid more aggressively in eBay auctions with red backgrounds (Bagchi & Cheema, 2012).
- **Evolution and Glossy Colors.** You prefer glossy colors because of the instinctive need for water: People who ate salty crackers preferred photos on glossy paper (Meert, Pandelaere, & Patrick, 2014). For the past year, I've been typing away at this book with a continuous supply of cashews on my desk. I had

assumed that my work ethic was causing my intense focus, but perhaps I'm fixated on my glossy screen merely because I'm thirsty.

- **Red and Sex.** On a real dating website—OkCupid—women were more likely to wear red in their photos if they indicated an interest in casual sex (Elliot & Pazda, 2012). Perhaps that's why men are more likely to pick up female hitchhikers who are wearing red (Guéguen, 2012).
- **Men and Red.** Men seem attractive in red, too. However, the reason stems from perceived dominance. Researchers tested this concept in four countries with different colors and men, and the results were the same: Women perceive a man in red to have a higher status, which—in turn—enhances his attractiveness (Elliot et al., 2010).

14

People

Figure 14A

LOOK AT THE BABY and number in Figure 14A. If you had to guess, which sex is the baby: male or female? In this chapter, you'll learn why the adjacent number biased your guess toward male.

Babies enter a sensory world filled with people, but what exactly is a person? How do you identify this concept?

I'd say four components:

- **Animacy.** Animate objects are alive. You often detect this animacy through motion—objects seem alive when they move in unexpected paths (Tremoulet & Feldman, 2000).
- **Agency.** Humans have goals and intentions. It's also called *theory of mind*.
- **Morphology.** Humans have a particular morphology, such as faces. Within 10 minutes of birth, infants orient their attention toward human faces (Goren, Sarty, & Wu, 1975).
- **Sex and Gender.** Many people confuse sex and gender. Sex involves anatomical parts, while gender involves an identification with the cultural traits of male or female.

This chapter might be the most complex topic in this book, but I'll try to simplify everything.

Let's start with gender.

What is Gender?

Reflect on your interactions with men and women. Do you believe that men and women are inherently different? That we behave in different ways?

If so, are these differences based on biological dispositions? Or do they emerge from psychological variables, such as culture?

Some researchers argue that very few differences exist, except for trivial motor skills (e.g., throwing distance; Hyde, 2005). However, differences exist because of the cultural stereotypes of gender. Every culture depicts "typical" traits for each sex, and these stereotypes nudge children to embody these traits.

I categorized this process into four stages:

1. Gender Labeling
2. Gender Scaffolding

3. Gender Reinforcement
4. Gender Identity

1. Gender Labeling

You learned that emotions—anger, fear, awe—aren't discrete biological concepts. You feel these emotions *because* of these terms.

Gender works in a similar way. The mere dichotomy of MALE and FEMALE created two containers that we filled with meaning.

How did we fill these concepts with meaning? Gender scaffolding.

2. Gender Scaffolding

Once children learn MALE and FEMALE, they start noticing cues and traits during their experiences.

> Young children search for cues about gender—who should or should not do a particular activity, who can play with whom, and why girls and boys are different. From a vast array of gendered cues in their social worlds, children quickly form an impressive constellation of gender cognitions (Martin & Ruble, 2004, p. 67).

Many children learn these differences through external sources, such as the media (e.g., books, TV, movies). And that's problematic.

As you'll see, children pursue their conceptual knowledge of gender. If children perceive themselves to be MALE, they will adopt whichever traits they inserted into their concept of MALE. Eventually, children will *become* these concepts. Media outlets are crucial because they often provide children with these typical traits. The media, right now, is dictating the type of people that children become in society. And these children become adults who get jobs in the media, perpetuating a vicious cycle for future generations.

If the media is so important, how is it currently portraying gender to children? Unfortunately, it's not great. Researchers analyzed elementary school textbooks, and they found pervasive stereotypes (Evans & Davies, 2000). Roughly 24% of the males were aggressive (compared to 4.9% of females), while 29.5% of females were passive (compared to 8.4% of males). Television isn't better: Among 467 commercials for children, females were less likely to appear in occupational roles (Davis, 2003).

Once we fix the media, then we'll be saved, right? Unfortunately, many stereotypes also emerge from gender assignments in language. These assignments are supposed to be arbitrary, but are they? English speakers are eerily accurate at guessing gender assignments in other languages (Boroditsky, Schmidt, & Phillips, 2003).

Gender assignments aren't random, which is problematic because of the omnipresence of language:

> . . . speakers of languages with grammatical gender must mark gender almost every time they utter a noun (hundreds or thousands of times a day). The sheer weight of repetition (of needing to refer to objects as masculine or feminine) may leave its semantic traces, making the objects' masculine or feminine qualities more salient in the representation (Phillips & Boroditsky, 2003, p. 929).

Whenever children say "doll" in Spanish—which has a feminine assignment—they insert this idea into their concept of FEMALE.

To recap, children build their gender concepts from external sources, like media and language. As you'll see next, we should also blame parents.

3. Gender Reinforcement

So far, children have (a) learned MALE and FEMALE, and (b) filled these concepts with meaning. At this point, they gravitate toward a particular side. And, oftentimes, parents are the source of that push.

Many parents nudge their children toward a particular side without realizing it:

That's my sweet little girl.
That's my big strong boy.

Children hear those remarks, and they insert these associations into their concepts of gender:

> Adults in the child's world rarely notice or remark upon how strong a little girl is becoming or how nurturant a little boy is becoming, despite their readiness to note precisely these attributes in the "appropriate" sex (Bem, 1981, p. 355).

These remarks seem trivial, but they have profound consequences in society. Boys and girls earn similar grades in math and science, yet girls are less likely to pursue careers in science, technology, engineering, and mathematics (Blickenstaff, 2005). Parents assume that their daughters are less interested in science, and they nudge them toward other career paths (Tenenbaum & Leaper, 2003). In order to fix these biases, we need to fix the stereotypes.

4. Gender Identity

After parents nudge their children toward one side of the gender dichotomy, children begin classifying themselves as this gender. And they become motivated to adopt the "typical" traits:

Many young children pass through a stage of gender appearance rigidity; girls insist on wearing dresses, often pink and frilly, whereas boys refuse to wear anything with a hint of femininity (Halim et al., 2014, p. 1091).

This behavior extends into adulthood. For example, participants played a video game that dropped bombs on other people. Participants who wore large nametags—thereby activating their gender concept—conformed to stereotypes: Males dropped more bombs, while females dropped fewer bombs. When everyone was truly anonymous, they dropped bombs equally (Lightdale & Prentice, 1994).

In sum, children enter a world that distinguishes a gender of MALE and FEMALE. Children learn the "typical" traits for each gender, usually through external sources (e.g., media, language). Next, parents subtly (or not-so-subtly) push them toward a particular side of the dichotomy, and these children adopt the traits they inserted into these genders.

Sex Differences

Men and women differ by cultural stereotypes of gender, but they also differ by sex and biology. Men have more testosterone, which can increase their physical aggression. Women are often more empathetic because they have less testosterone (Hines, 2011; Chapman et al., 2006; Pasterski et al., 2007; Baron-Cohen & Wheelwright, 2004).

Evolutionary circumstances play a role, too. Choose the option that feels worse:

▶ **Option A.** Your partner enjoys passionate sex with somebody else.

▶ **Option B.** Your partner forms a deep emotional connection with somebody else.

Most men choose Option A, while most women choose Option B (Buss, Larsen, Westen, & Semmelroth, 1992). These two types of jealousy were adaptive for our ancestors:

▶ **Sexual Jealousy.** Male ancestors were more likely to reproduce if they felt sexual jealousy. If a female became pregnant with another man's child, the original male lost his ability to reproduce (as well as paternal certainty). Males who felt sexual jealousy were more likely to stop the infidelity and successfully reproduce, thereby passing on a disposition toward sexual jealousy.
▶ **Emotional Jealousy.** Female ancestors were more likely to reproduce if they felt emotional jealousy. If a male became emotionally attached to another woman, the original female lost her ability to acquire resources. Females who felt emotional jealousy were more likely to stop the infidelity and acquire life-nourishing resources. Her children were more likely to survive until reproduction, thereby passing on a disposition toward emotional jealousy.

Other sex differences exist, too. Women, by nature, invest more time and energy into reproduction: pregnancy, birth, breast-feeding, and the list goes on. It's not easy. Women won't have children with any ol' schlum. Men, on the other hand, have it easy. What do we do? Nothing. Just the fun part.

Thus, women prioritize quality over quantity:

> ... because members of the slow sex are limited to a relatively small number of offspring, they suffer a relatively great reproductive loss if any of these offspring fail. For

these reasons women are expected to evolve traits that maximize offspring quality rather than quantity (Bailey, Gaulin, Agyei, & Gladue, 1994, p. 1081).

Today, women are more selective in choosing mates. Researchers gave participants a list of 24 traits (e.g., intelligent, kind, friendly), and they asked people to indicate the minimum percentile of each trait that would be acceptable in order to have a one-night stand with somebody. Men set lower demands across all 24 dimensions (Kenrick, Groth, Trost, & Sadalla, 1993).

Women can usually find a man who wants to have sex, so they can reproduce more easily than men. Some researchers argue that parents favor daughters during economic downturns because these daughters can find a mate (and produce grandchildren) more easily:

> . . . as resources become increasingly scarce, parents should view their female children as the safer reproductive option because doing so increases the probability of avoiding the catastrophic evolutionary outcome of producing zero grandchildren (Durante, Griskevicius, Redden, & Edward White, 2015, p. 4).

That finding is pretty interesting, but take heed. Later, I'll explain why I disagree with the conclusion. Can you spot another explanation?

Anthropomorphism

Over time, you start inserting human traits—animacy, agency, morphology, sex, gender—into nonhuman objects.

When you see cars, for example, you immediately look at

headlights. Why? Because they resemble human eyes (Windhager et al., 2010). This attribution of human traits is called *anthropomorphism*.

You anthropomorphize for two reasons:

- ▶ **Sociality Motivation.** You desire a social connection.
- ▶ **Effectance Motivation.** You want to understand something.

First, you attribute human traits if you crave a social connection. Researchers displayed a continuum of images that progressed from doll to human, and they asked people to indicate the point at which the doll became alive. People who felt socially disconnected lowered their threshold (Powers, Worsham, Freeman, Wheatley, & Heatherton, 2014).

Interestingly, you also anthropomorphize in cold temperatures. People who drank cold tea generated more anthropomorphic ideas for a robot, such as talking or walking (Lee, Rotman, & Perkins, 2014). Why? People who are seeking physical warmth will unknowingly be seeking social warmth.

Unfortunately, this effect has a reciprocal (and dangerous) effect: Anthropomorphism can satisfy your desire for human contact. When people viewed an anthropomorphic design of a Roomba vacuum, they planned to spend less time talking with friends and family that week because the vacuum satisfied their need for human contact (Mourey, Olson, & Yoon, 2017).

Today, humans have a proliferation of smart devices, voice assistants, and anthropomorphic gadgets. Yet, despite the era of social media, humans are feeling lonelier than ever (Cacioppo & Patrick, 2008). Perhaps we could blame the heightened anthropomorphism. Humans might be satisfying their desire for sociality through devices, yet this behavior isn't solving our true need for human contact. As a society, we should reconsider the long-term implications of our technology so that we don't replace genuine interactions.

We never touched upon effectance motivation, but we'll revisit this concept in a later chapter.

Warmth and Competence

Our ancestors felt an incessant desire for survival, so they developed an ability to evaluate people (and threats) very quickly. Today, you can judge people within 39 ms of seeing their face (Todorov, Pakrashi, & Oosterhof, 2009).

Those judgments emerged from an evolutionary foundation, so they typically involve survival traits: People judge trustworthiness faster than they judge intelligence (Bar et al., 2006).

Two traits were most important:

- **Warmth.** Does somebody intend to help or hurt us?
- **Competence.** Can this person fulfill those intentions?

You typically correlate these traits for individuals: If somebody has warmth, you attribute competence; if somebody has competence, you attribute warmth.

However, you reverse those correlations for groups:

- **Age.** Seniors are warm; the youth are competent.
- **Wealth.** The poor are warm; the rich are competent.
- **Gender.** Females are warm; males are competent.

These associations can lead to harmful stereotypes: Males seem less warm, while females seem less competent.

Remember the study in which parents favored their daughters during an economic downturn, presumably because of their ability to produce grandchildren? I disagree. Parents favored their daughters because they seemed less competent, as if they'd be less capable of handling the crisis. Researchers could test that explanation by examining this effect with daughters who (a) don't want children, or (b) can't have children. Parents should have no reason to favor those daughters because they won't produce grandchildren, but I suspect they *will* favor them for the lack of competence.

Finally, let's combine warmth and competence with our tendency to anthropomorphize. Evidence shows that hurricanes with female names kill significantly more people (Jung, Shavitt, Viswanathan, & Hilbe, 2014). Why? You attribute the human trait of FEMALE (including warmth) to these hurricanes. They seem less severe, so people don't evacuate. In a lab study, people perceived male hurricanes (e.g., Arthur, Kyle, Marco) to be more severe than female hurricanes (e.g., Bertha, Laura, Hanna; Jung et al., 2014).

In fact, the dichotomy of MALE and FEMALE is so pervasive that you attribute gender to *any* dichotomy. Every time that you categorize something into two entities, you conceptualize these entities with human gender.

You can see this effect by creating a dichotomy *within* females, such as housewives and feminists:

> On the one hand, traditional women (e.g., housewives) may be judged to be warm but not particularly competent... On the other hand, nontraditional women (e.g., feminists) may be seen as competent but hostile (Judd, James-Hawkins, Yzerbyt, & Kashima, 2005, p. 900).

Even though this dichotomy appears within females, you conceptualize these two groups as MALE and FEMALE.

But again, it's not just people—you attribute gender to *any* dichotomy. Researchers asked people to evaluate two versions of a website:

- **Profit:** www.mozilla.com
- **Nonprofit:** www.mozilla.org

The profit company seemed masculine and competent, yet the nonprofit company seemed feminine and warm (Aaker, Vohs, & Mogilner, 2010). You confuse any dichotomy for the dichotomy of human gender.

Numbers

Numbers have dichotomies, too. They can be odd or even.

Remember the baby image at the beginning of the chapter? Babies seemed masculine near an odd number, yet they seemed feminine near an even number. Other studies found similar results with different methodologies (see Wilkie & Bodenhausen, 2015).

Same with the dichotomy of precision: Sharp numbers seem masculine, while round numbers seem feminine (Yan, 2016).

Phonemes

Compare the following groups of words. Which group seems masculine? And which is feminine?

- **Group 1.** Panyin, Sepast, Fopaz
- **Group 2.** Banyin, Zepast, Vopaz

Most people believe that Group 1 is feminine, whereas Group 2 is masculine. Is that what you chose?

First, you should understand that we associate gender with rigidity. In one study, researchers asked people to guess the gender of a face that was midway between male and female. Participants saw a male face while squeezing a rigid ball, yet they saw a female face while squeezing a foam ball (Slepian, Weisbuch, Rule, & Ambady, 2011).

How does rigidity involve phonemes? Voiced phonemes (e.g., b, d, g) sound HARD because their vibrations produce a harsher sound in the vocal cords. Unvoiced phonemes (e.g., f, s, t) sound SOFT because they produce a breathier sound. Therefore, participants believed that Group 1 (Panyin, Sepast, Fopaz) was feminine, yet Group 2 (Banyin, Zepast, Vopaz) was masculine (Slepian & Galinsky, 2016).

Projective Simulation

Your perception is influenced by anchoring. If you see a number, say 37,465, you'll estimate higher numbers in subsequent tasks.

For example, how many paragraphs are in this book? Got your guess? The true answer is near 4,000, but your guess is now higher than it would have been because of your exposure to 37,465. While estimating the number of paragraphs, you started at this anchor point and then adjusted downward until you reached a reasonable estimate. A lower anchor, say 14, would have nudged you to adjust upward—and your new estimate would be lower.

Anchoring is the key mechanism behind an important concept with people. In Chapter 2, you saw images in which I was extending my arm, and you immersed yourself into my body. You perform these simulations to decipher the inner thoughts of other people, an ability called mentalizing. Uploading pictures to Facebook? You might simulate how other people (e.g., boss, ex-lover, new crush) would react to your pictures.

I described that principle as *agent simulation*, but I'll use the term *projective simulation* because it reinforces the key idea: You immerse *yourself* into an agent. You judge other people's mindsets by adjusting away from *your* mindset (Nickerson, 1999).

For instance, participants listened to a voicemail that was ambiguously sarcastic, and they judged whether other people would interpret the message as sarcastic. When participants were told the voicemail *was* sarcastic, they estimated that other people would recognize the sarcasm. They projected their private knowledge into those people (Epley, Morewedge, & Keysar, 2004).

Both children and adults suffer from that bias: Adults adjust (insufficiently) from their perspective, while children fail to adjust altogether (they assume that everybody shares their private beliefs; Birch & Bloom, 2004).

Canonical Gender

Canonical prototypes are the "typical" traits of an object. They are basically stereotypes, so they become problematic when dealing with people.

Read this passage:

> This morning a father and his son were driving along the motorway to work, when they were involved in a horrible accident. The father was killed and the son was quickly driven to the hospital severely injured. When the boy was taken into the hospital a passing surgeon exclaimed: "Oh my god, that is my son!"(Reynolds, Garnham, & Oakhill, 2006, p. 890).

Many people are confused by that passage. Were you?

Even though I've been discussing gender biases, people still have trouble simulating a surgeon as a female (e.g., mother). People are confused by that passage because males are the canonical gender for surgeons.

Other occupations have canonical genders, too. In one study, people read occupations that were stereotypically male (e.g., executives, sheriff, hunter) or female (e.g., secretary, florist, cheerleader). They were slower to read subsequent pronouns that were incongruent (e.g., "she" after "executive") because they were simulating the opposite gender (Kennison & Trofe, 2003).

Those implications can be detrimental in society. While hiring a new employee, managers will simulate a canonical prototype of the candidate (e.g., personality, appearance, gender), and they will prefer candidates who match this prototype.

Remember how women avoid careers in science and technology? These occupations have a canonical prototype of MALE. Since managers prefer candidates who match this prototype, they are more likely to hire men. Researchers submitted fictional job applications to science departments at universities, and they alternated a male and female name (while everything else remained the same). Those hiring managers—including women—were more likely to hire applicants with a male name, and they offered male applicants a higher salary ($30,238 for males vs. $26,508 for females; Moss-Racusin, Dovidio, Brescoll, Graham, & Handelsman, 2012).

You have canonical races, too. Researchers sent fictional resumes throughout Boston and Chicago; applicants with stereotypically White names received 50% more callbacks (Bertrand & Mullainathan, 2004).

Despite our implicit biases, most people strive to be fair; they simply fail to notice these biases. Later, I'll offer solutions to combat these issues in society.

So here we are . . . we finally tackled the primitive concepts. Until now, we've seen mostly quirky examples. But who cares if we associate males with odd numbers? Why is that important?

It's not important. However, the next part of the book will explain how these concepts guide your perception and behavior on a much larger scale. It's a bit scary. Yet, if we understand these concepts, we can start controlling them. We can improve our lives. Improve society. I see a lot of potential for these concepts, and I'm excited for the road ahead.

Summary

Humans possess specific traits—animacy, agency, morphology, sex, and gender. You start attributing these traits to inanimate objects, especially when you (a) desire a social connection, or (b) want to understand something.

In particular, you confuse any dichotomy for the dichotomy of human gender, inserting the traits of gender (e.g., warmth, competence) into those ideas.

You also engage in projective simulation—you decipher the inner thoughts of other people by adjusting away from your mindset.

Other examples:

- **Sociality Motivation.** If somebody denies your friend request on social media, you become more likely to buy products with anthropomorphic spokespeople because you desire a social connection (Chen, Wan, & Levy, 2017).
- **Spotlight Effect.** If you simulate yourself into other people, where would you look? Probably at yourself, right? When you're in public, you falsely believe that everybody is aware of—and perhaps judging—your behavior (when that's not actually the case). Researchers asked people to walk into a room wearing an embarrassing t-shirt (e.g., a large picture of Barry Manilow), and they vastly overestimated the number of people who actually noticed the shirt (Gilovich et al., 2000). People are not looking at you as much as you believe they are.
- **Gender Assignments.** Hurricanes seem more severe with a masculine name. Language produces a similar effect with gender assignments: German speakers perceive the sun to be warm and nourishing because of its feminine assignment, while Spanish speakers perceive the sun to be powerful and threatening because its masculine assignment (Boroditsky

et al., 2003). Perhaps these differences are influencing the perceived threat of climate change: Societies with masculine assignments for the sun might work harder to combat global warming because it seems more powerful and threatening.

▶ **Canonical Location of Gender.** Male names typically appear before female names (e.g., Adam and Eve), which reinforced a canonical location of gender: MALE is LEFT, while FEMALE is RIGHT. But again, it depends on reading directionality: Italian speakers drew men on the left, while Arabic speakers drew men on the right (Maas et al., 2009).

▶ **Canonical Traits in Hiring.** Biases occur with *any* canonical trait. Managers are less likely to hire male applicants for stereotypically feminine jobs (e.g., secretaries), and they are less likely to hire attractive people for unattractive jobs (e.g., truck driver; Glick, Zion, & Nelson, 1988; Johnson, Podratz, Dipboye, & Gibbons, 2010). People aren't necessarily advantaged because they are white or male or beautiful; they are advantaged if they match the canonical prototype.

1
Origin

2
Primitives

3
Applications

In this part, you'll see how primitive concepts shape your perception and behavior on a broader scale.

15

Beauty

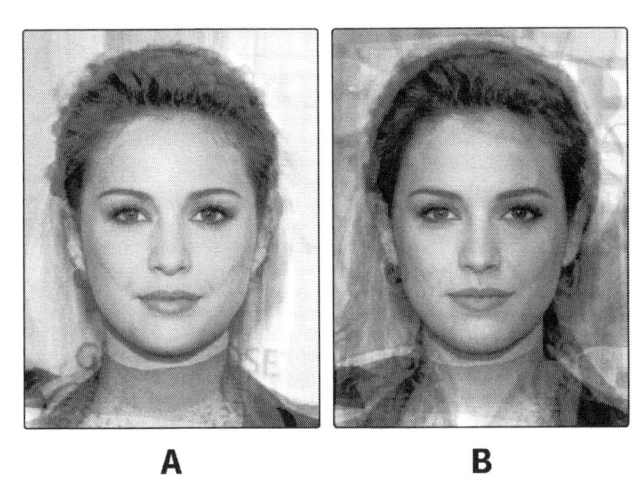

A   **B**

Figure 15A

WHICH FACE IS more attractive in Figure 15A?

Most people prefer Version B. Did you? If so, why? Both pictures are nearly identical, so what features did you prefer? What traits? Anything? In this chapter, you'll discover the hidden elements that are constructing your perception of beauty.

Canonical Faces

You encounter many faces across your life, and you merge these faces into an average face—that is, a *canonical face*.

You prefer faces that match this prototypical face. People are more attracted to Version B in Figure 15A because that face is closer to the population average.

Why is Version B an "average" face? Because this photo is a digitized combination of *many* faces. Researchers find a linear trend: More faces become more attractive (Langlois & Roggman, 1990).

You can see a clear example in Figure 15B. Male faces start with Justin Bieber and Chris Pratte, whereas female faces start with Selena Gomez and Miley Cyrus (see Face Facts, 2019).

You typically assume that beautiful faces are distinctive and

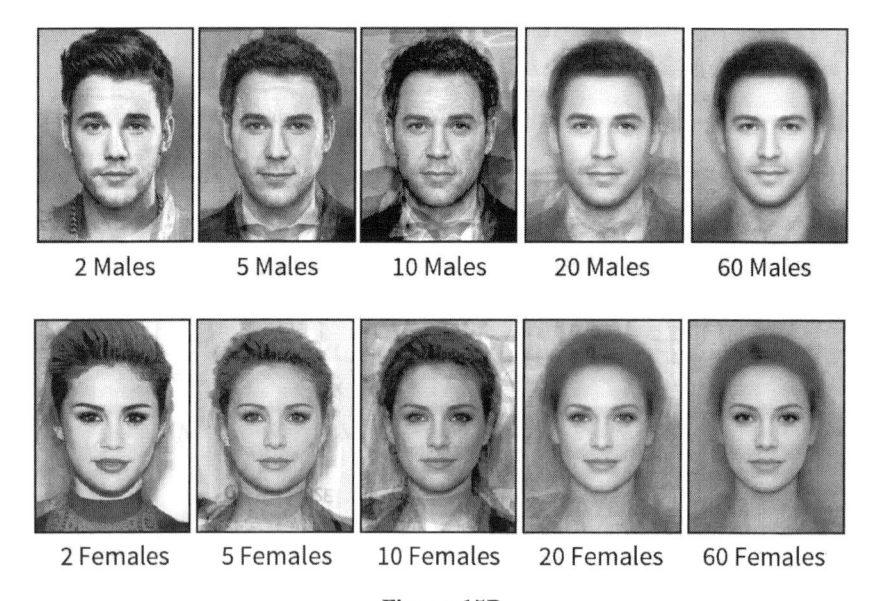

Figure 15B

noteworthy, but I'm arguing the opposite: Beautiful faces are typical and average. Why is that? Researchers argue that averageness helped our ancestors find mates:

> If we had an innate template for attractiveness, we could run the danger of either never meeting somebody fitting the template, or being frustrated by the non-fitting mates we find . . . there are simply more average people than extremes. This creates a bigger population of possible mates (Grammer, Fink, Møller, & Thornhill, 2003, p. 394).

Averageness helped our ancestors find mates and reproduce, gifting their offspring with this same preference.

Average faces are also congruent with your prototypical face, activating positive emotions. And you rationalize these feelings by generating hypotheses: *Hmm, something about this face feels right. Therefore, I must be attracted to it.*

More on this later.

Components of Canonical Faces

How many faces have you seen across your life? Millions? Tens of millions? The number is so obnoxiously large that I can't fathom a ballpark estimate.

So then, why don't we all possess a similar prototype? I prefer brunettes, but other people prefer blondes. Why are we attracted to different "types" of people?

My guess: We build a canonical face in childhood, and we adjust this prototype based on the people we see most often and recently.

That statement has three components of a canonical face: early experience, frequency, and recency.

1. Early Experience. I need to share disturbing news. Ready? You are physically attracted to your parents. Hear me out.

If you build a canonical face during childhood, then your prototype will contain faces that you see most often during this period, such as your parents. And you will be attracted to these faces.

Indeed, it seems that way:

- ▶ Women are more likely to date men with the same eye color as their father (Wilson & Barrett, 1987), and they gravitate toward younger or older men in accordance with the age of their parents (Perrett et al., 2002).
- ▶ When people were subliminally exposed to an image of their parents, they found a subsequent photo to be more sexually attractive (Fraley & Marks, 2010).
- ▶ While viewing images of strangers, participants could match wives with their mothers-in-law above chance (Bereczkei, Gyuris, Koves, & Bernath, 2002)

Luckily, you developed evolutionary disgust to counteract that attraction; otherwise you would produce unhealthy offspring.

2. Frequency. You build the primary template in childhood, but you adjust this canonical face later in life, toward the faces that you see frequently.

Every day, you see your own face in the mirror. So then, wouldn't your prototype contain your own face? Wouldn't you be attracted to your face? It sounds absurd, but it's true. Researchers gave participants multiple photos of their romantic partner, and they asked people to choose the most attractive photo. Sure enough, they chose photos that had been secretly morphed with 22% of their own face (Laeng, Vermeer, & Sulutvedt, 2013).

You gravitate toward people who look like you. It's no coincidence that many people look like their pets. Pets don't magically alter their face to conform to their owner; instead, people choose pets that resemble them.

Heck, it even happens with cars: Participants could match the fronts of cars with respective owners (Stieger & Voracek, 2014).

3. Recency. You also adjust your canonical face toward recent faces. After seeing blended faces containing humans and chimps, people became more attracted to human faces with subtle blends of chimp (Principe & Langlois, 2012).

Media outlets can be psychologically detrimental because they often display a high quantity of unusually attractive people. These exposures can shift your canonical face away from the true average, and you start finding other people—such as your romantic partner—less attractive (Kenrick & Gutierres, 1980). In order to combat this detriment, you need see a wide variety of faces to construct a truly average prototype.

Why the Beauty Bias is Flawed

You believe that attractive people are superior on many dimensions:

- ▶ Intelligence (Jackson, Hunter, & Hodge, 1995)
- ▶ Sociability (Feingold, 1992)
- ▶ Trustworthiness (Shinners, 2009; Wilson & Eckel, 2006)

You behave differently, too: Managers hire attractive people . . . judges give lenient sentences to attractive criminals . . . people vote for attractive politicians (Bóo, Rossi, & Urzúa, 2013; Sigall & Ostrove, 1975; Todorov, Mandisodza, Goren, & Hall, 2005).

Researchers argue that beauty *causes* those attributions: *Hmm, this person is attractive. They must be smart.*

However, I think the real culprit is congruence. Whenever

a stimulus matches your canonical prototype, you feel positive emotions. And you misattribute these emotions: *Hmm, something about this stimulus feels good. It must be [insert a reason].*

I'm arguing that beauty is merely another hypothesis. In other words, here's the old explanation of the beauty bias: *Hmm, this person is attractive. Therefore, they must be [smart, sociable, trustworthy].* But I'm arguing that attractiveness belongs in the second portion of that statement: *Hmm, something about this person feels right. Therefore, this person must be [attractive, smart, sociable, trustworthy].*

Ultimately, beauty isn't the cause of the beauty bias. Beauty is merely another attribution. The true culprit is congruence. You'll see why this distinction is important soon.

Why is Symmetry Beautiful?

Symmetrical faces are more attractive, supposedly because they seem healthier (Rhodes et al., 2001).

But I disagree. In order to see the true reason, you should recognize that everything in the world is encapsulated with symmetry.

> All the forces we know in nature arise from principles of symmetry... it may be that a truly fundamental theory will be simply encapsulated in one simple statement of a symmetry principle (Steven Weinberg, Nobel Laureate in Physics, interview).

In fact—as we speak—nature is growing more symmetrical. Evolution is occurring beyond humans and animals; nature is also evolving toward forms and shapes that transport energy more effectively. And those structures are often symmetrical (see Figure 15C; Bejan & Lorente, 2010).

Lightning   Rivers   Lungs

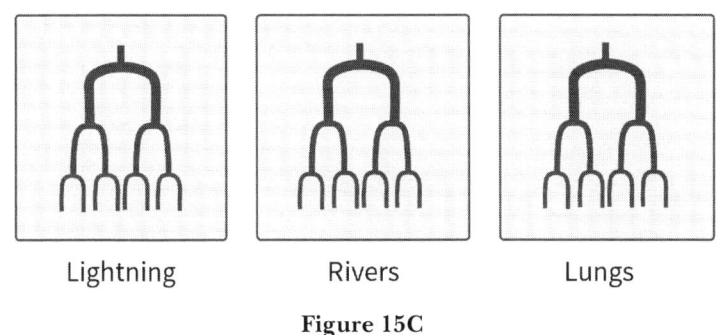

Figure 15C

Similar effects happen with faces. Faces have an underlying blueprint of perfect symmetry. No face will have *perfect* symmetry, but the quintessential average face *does* have perfect symmetry.

Sound familiar?

You learned that average faces are attractive.

And now you learned that average faces are symmetrical.

You are attracted to symmetrical faces *because* those faces are average. So then, why does evidence indicate that symmetrical faces seem healthier?

Once again, you can blame congruence: Average faces will seem favorable on *any* trait. Much like researchers have misdiagnosed beauty as the cause of the beauty bias, I believe they have misdiagnosed healthiness as the cause of symmetrical beauty. In reality, healthiness is merely another attribution: *Hmm something about this person feels right. They must be [healthy, attractive, smart, sociable, trustworthy]*.

Why Do You Prefer Certain Traits?

Most women prefer taller and older men, whereas most men prefer shorter and younger women. Let's find out why.

Height

Most women seek a man with power, presumably because our female ancestors sought this trait. Consequently, we assume that women prefer tall men *because* these men seem powerful.

But what if this relationship is reversed? What if power isn't the cause? What if power is the effect?

Women are typically shorter than men. Naturally, they become attracted to tall men because this upward angle matches their canonical perspective of men.

Women desire UP. And, as you recall, UP is POWER.

Therefore, women who desire UP unknowingly desire POWER. So, perhaps women don't desire POWER (and thus tall men). Perhaps they desire tall men (and thus POWER).

Side question: Could these viewpoints perpetuate sexism? Since men are taller, they literally "look down" at women. Could this viewpoint instill a metaphor of condescension?

Age

Women prefer older men, while men prefer younger women. Personally, I was confused by those preferences. If a woman is 25 years old, she might prefer a 28-year-old. If she is 30 years old, she might prefer a 33-year-old. In both cases, she prefers a man who is a few years older than her current age. But why? If she desires POWER, wouldn't she prefer the most powerful male—in this case, the 33-year-old—at *any* age? Why does she prefer ages that are *relative* to her age?

First, you should understand that SIZE involves a relative comparison: You evaluate sizes through a "more" vs. "less" judgment.

Therefore, women insert this relative trait of SIZE into their prototype for men: They notice that men are "slightly" taller, so their concept of MALE inherits a magnitude that is slightly larger than themselves.

Do you see the issue?

You insert SIZE into all abstract forms of size, such as height and age. And you confuse these sizes. Women prefer men who are slightly taller because of their canonical perspective, and they start preferring men who are slightly older because both magnitudes are pulling from the same primitive ingredient of SIZE.

It's the same effect you saw in Chapter 6. You believe that you have less in common with a store clerk who is geographically far away because you confuse the spatial distance for social distance.

Let's walk through an example. In the US, a typical height for women is 160 centimeters, while a typical height for men is 175 centimeters. For these women, men are roughly 10% taller. If these women confuse height magnitude with age magnitude, then they will prefer men who are roughly 10% older. If a woman is 25 years old, her male prototype will be 27.5 years old. If she is 30 years old, her male prototype will be 33 years old. Ultimately, women prefer men who are slightly older *because* they prefer men who are slightly taller. They confuse height magnitude for age magnitude.

If my speculation is true, then taller women might prefer younger men, whereas shorter men might prefer older women.

Facial Recognition

Canonical faces also help you *recognize* faces. Consider two people who look similar. How would you distinguish them? Wouldn't you need to identify minor differences in their face? Indeed, your canonical face needs a variety of building blocks to notice these subtle distinctions.

So, who is better at recognizing faces?

- ▶ Someone raised in Boston, Massachusetts
- ▶ Someone raised in Amidon, North Dakota

Big cities, like Boston, have many people. How many faces do residents typically see? Quite a lot, right? Naturally, they acquire more facial features in their working prototype. However, residents of small towns—like Amidon, with a population of 20 people—are exposed to fewer faces, acquiring fewer building blocks in their canonical face. Not surprisingly, people raised in big cities, like Boston, are superior at recognizing faces (see Balas & Saville, 2015).

Early experience is also crucial. People who are born blind, yet gain sight later, never recognize faces with the same aptitude as other people. In fact, since the left eye passes visual input to the brain's right hemisphere, which controls facial processing, this deficit occurs for people who are only blind in their left eye (Mondloch, Le Grand, & Maurer, 2003).

The types of faces matter, too. Researchers noticed that infants see a female face of the same race (usually the mother) more often than other faces (Rennels, & Davis, 2008). These exposures dictate your perception later in life.

For one, you experience a *female recognition bias*: You can process female faces more easily (Ramsey-Rennels & Langlois, 2006).

You also experience an *own race bias*: You can distinguish faces of your own race more easily (Meissner & Brigham, 2001). But that terminology is slightly inaccurate. You aren't biased toward your own race; you are biased toward the race that you see most often in childhood.

Researchers studied three groups of people:

- ▶ Caucasian infants in a Caucasian environment
- ▶ African infants in an African environment
- ▶ African infants in a Caucasian environment

The third group didn't show that bias (Bar-Haim, Ziv, Lamy, & Hodes, 2006).

We need to give children many varied (and positive) exposures to different races. Children who see many different races will build a canonical face with more diversity. Later in life, they will possess flexible prototypes with different skin tones, structures, and other traits. Many races will be congruent with their canonical face.

You find a similar mechanism for recognizing emotions. If an angry father continuously abuses his daughter, she'll construct a canonical face that is angrier. When researchers displayed a spectrum of angry faces that progressed from fuzzy to clear, physically abused children could identify angry faces faster than other children (Pollak & Sinha, 2002). Abused children needed fewer details to recognize an angry face because this face resembled their canonical prototype.

I'm only speculating, but this mechanism could also explain why some people gravitate toward abusive partners. An abused daughter might construct a prototype of MALE that consists of anger and abuse, and she will seek partners who are congruent with this prototype. If you emerged from a healthy childhood, you might be quick to criticize people who seek unhealthy traits in a partner, but hopefully you see the deep-rooted reasons behind that behavior.

Summary

You develop a canonical face, an average face based on the people you typically see. You are attracted to these average faces because you misattribute this pleasant sensation: *Hmm, something about this person feels right. I must find this person attractive.*

Your canonical face also helps you recognize faces more easily. Children gain more proficiency at facial recognition, especially for other races, when they see a wide diversity of faces.

Other examples:

- ▶ **Geography.** People with close ties (e.g., spouses, siblings, friends) share similar preferences for attractiveness because they reside in the same community, and they see the same faces (Bronstad & Russell, 2007).
- ▶ **Species Bias.** Children become proficient at distinguishing monkey faces if they are exposed to many monkey faces (Pascalis et al., 2005).
- ▶ **Color Averaging.** You merge all faces into an average face. Perhaps the same effect occurs with color, which might explain why WHITE is GOOD (Kareklas, Brunel, & Coulter, 2014). When you mix lights of all possible colors, the end result is white. Since you are engulfed in a sea of color, perhaps WHITE is GOOD because WHITE is the average color of the world around you.

16

Morality

ARE HUMANS MORAL? Like *actually* moral in our nature? Or do societies need legal consequences to stop us from killing each other?

You might assume that we need external pressures because of our selfish Darwinian nature. And I think so too. Yet, at the same time, we do care for other people. In this chapter, I'll explain (a) how we derived these morals and (b) how society can apply that knowledge to increase morality.

The Origin of Empathy

Look around you. You might see discrete objects—books, pens, trees—but you are surrounded by a continuum of stimuli. You can perceive *anything* to be an object, as long as this entity adheres to Gestalt principles (e.g., proximity, similarity, containment).

For example, I'm holding a pen. So, how many objects am I holding? One? What if I remove the cap? Am I still holding one object? Or two objects? Again, objects have flexible boundaries—it just depends on how your brain is perceiving these objects.

Grouping can trick your brain, too. Next time that you open your freezer, you might see a container of ice cream next to frozen

196   THE TANGLED MIND

vegetables. If you perceive these two objects as a single group, perhaps because they are close together, the ice cream will seem healthier because it inherits healthiness from the vegetables. See Chapter 4 to refresh your memory of convergent processing.

This chapter will argue that grouping was a primary factor behind the evolution of human empathy, which has been a puzzling mystery in psychology research. Evolution favored people who were selfish, right? So, why do humans care for other people? Why would we sacrifice our own welfare to enrich the welfare of somebody else?

Grouping provides an answer.

Until now, grouping has involved inanimate objects—but this mechanism also happens with people (see Figure 16A).

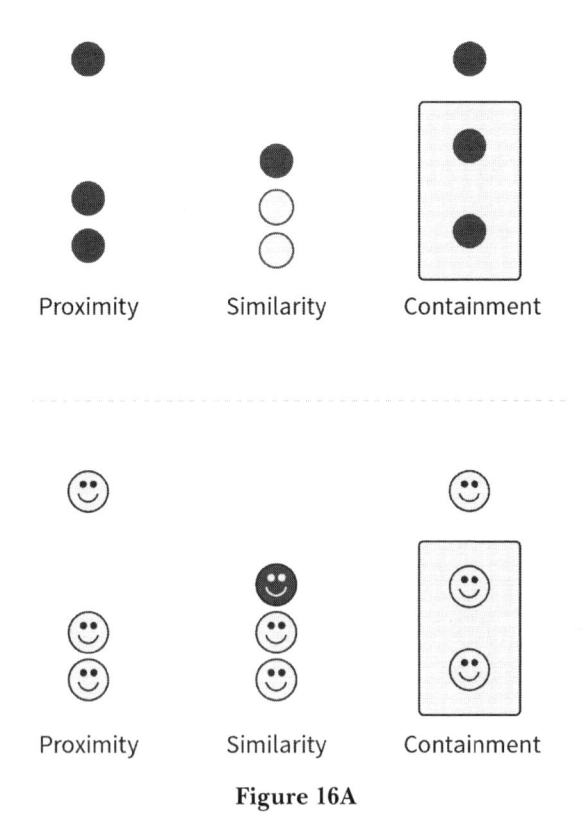

Figure 16A

Gestalt principles can group you with another person, merging both of you into a single object. You care for this person because you *are* this person:

> . . . when one takes the perspective of another (either through instructions or a feeling of attachment) and vicariously experiences what the other is experiencing, one comes to incorporate the self within the boundaries of the other . . . if the distinction between the self and other is compromised by perspective taking, then so is the distinction between selflessness and selfishness (Cialdini, Brown, Lewis, Luce, & Neuberg, 1997, p. 482).

The next few sections will highlight some striking coincidences between empathy and principles of grouping.

Proximity

You group stimuli that are close together.

However, you don't need to be physically close to somebody. You insert DISTANCE into many abstract forms.

Spatial Proximity. Somebody has probably died while you've been reading this chapter, but you probably don't feel any emotion right now. Why? Because that person is distant from you. When somebody dies across the world, you don't blink an eye. But if somebody dies across the street, you feel more emotion—even if you've never met either individual—because you are closer to this person.

Temporal Proximity. Are you still mourning the death of Socrates? How about Galileo? Probably not. They existed far away from you in time, which is built with DISTANCE.

Social Proximity. You also insert DISTANCE into social relationships: You show more empathy toward significant others because you are "close" to these people.

Similarity

You group stimuli that look similar.
   Therefore, you will group yourself with people who resemble you in visual or conceptual ways.

Race. Unfortunately, skin color influences grouping. In 2005, Hurricane Katrina struck the US with deadly force. When researchers asked people to infer the emotions of victims, people attributed fewer secondary emotions (e.g., anguish, mourning, remorse) to other races (Cuddy, Rock, & Norton, 2007).

Morphology. Human morphology might be the strongest similarity: You group with other humanlike forms. For example, you empathize with animals across this sequence: invertebrates, fish, amphibians, reptiles, birds, mammals, primates (Harrison & Hall, 2010).

Identifiability. You feel empathy with humanized depictions. Participants were more likely to donate $5 to help an African girl name Rokia, compared to the same request to help "millions of people" (Small & Loewenstein, 2003). Despite millions of people, you can't group with a statistic. In another study, doctors showed more empathy toward patients when X-rays contained a photo of the patient (Turner & Hadas-Halpern, 2008).

Conceptual Similarities. You are more likely to help somebody who shares a first name or birthday (Burger et al., 2004). Similarities are powerful tools of persuasion because you group yourself with these people.

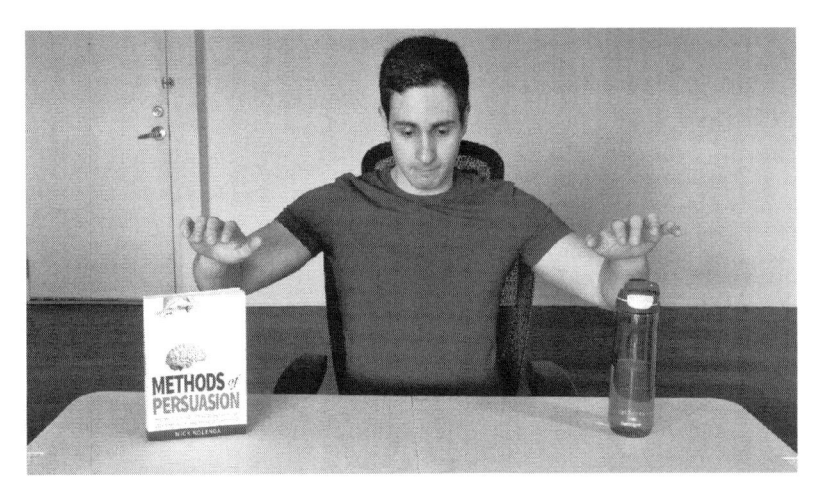

Figure 16B

Containment

You group stimuli inside the same enclosures.

But you don't need to jump inside a tub with somebody. You insert CONTAINMENT into abstract domains, such as ingroups.

Ingroups. You help injured strangers if they are wearing the clothes from your favorite sports team (Levine, Prosser, Evans, & Reicher, 2005). You help people who reside inside the same social group because you paint this imagery with physical containment. Both of you reside inside the same container.

Common Motion

You group stimuli that move together.

In Figure 16B, I'm sitting at a table with a book and water bottle. Suppose that I grab each object with my closest hand. Answer quickly . . . which hand grabs the book?

The correct answer is my right hand, but you were more likely to say my left hand if you feel connected to me. In one study, two participants sat across from each other, while one person guided the other person through a maze. If they belonged to different social groups, they performed better because they prevented their own perspective from biasing their instructions (Todd, Hanko, Galinsky, & Mussweiler, 2011). If you feel connected to me, you were more likely to group our bodies as a single unit. Moving your left hand would move my hand over the book because we're the same entity.

Chameleon Effect. Humans exhibit a *chameleon effect*: You nonconsciously mimic the gestures and movements of other people. You like people who mimic you, and you mimic people who you like (Chartrand & Bargh, 1999).

Well, mimicry is common motion. Both bodies move in tandem, forming a cohesive unit. You prefer people who mimic your behavior because you *are* them. Researchers could test that explanation by examining people with low self-esteem—those people might *dislike* mimickers because they would impute them with their low self-worth.

In sum, human empathy originated from a primitive tendency to group ourselves with other people. Humans are altruistic *because* of our selfish nature. Should the ulterior motive matter? Perhaps. Or perhaps not. Either way, we can start cultivating more empathy in society, regardless of the deeper motive.

All major religions have a "golden rule" to treat other people the way you want to be treated (see Figure 16C). Perhaps this doctrine is more literal than we thought.

Moral Scaffolding

Human morality also emerged from scaffolding. You developed your moral compass from primitive concepts, such as emotion.

Buddhism	Just as I am so are they, just as they are so am I', Sutta Nipata 705
Confucianism	Never impose on others what you would not choose for yourself', Analects XV:24
Hinduism	One should never do that to another which one regards as injurious to one's own self. This, in brief, is the rule of dharma' Brihaspati, Mahabharata (Anusasana Parva, Section CXIII, Verse 8)
Taoism	Regard your neighbor's gain as your own gain, and your neighbor's loss as your own loss', Tai Shang Ying Pian, Chapter 4
Judaism	That which is hateful to you, do not do to your fellow. That is the whole Torah; the rest is the explanation', Talmud, m. Shabbat 31a
Jainism	Just as sorrow or pain is not desirable to you, so it is to all which breathe, exist, live or have any essence of life', Jain sutra 155–156
Chrisitianty	Therefore all things whatsoever would that men should do to you, do ye even so to them', Matthew 7:12, Luke 6:31
Islam	As you would have people do to you, do to them; and what you dislike to be done to you, don't do to them', Kitab al-Kafi, Vol. 2, p. 146

Figure 16C. Every major religion has a "golden rule." Table adapted from Baumard & Boyer (2013a).

Read the following scenarios:

▶ A woman is cleaning her closet, and she finds an old flag. She doesn't want the flag anymore, so she cuts it up and uses the rags to clean her bathroom.
▶ A family's dog was killed by a car in front of the house. They heard that dog meat was delicious, so they cut up the dog's body, cooked it, and ate it for dinner.
▶ A brother and sister like to kiss each other on the mouth. When nobody is around, they find a secret hiding place and kiss each other on the mouth, passionately.

Those actions don't hurt anybody. So, are they moral? If not, why are they immoral? What's the reason?

Most college students in the Western world don't perceive any moral wrongdoings, but some participants in other countries (e.g., Brazil) often perceive moral violations, even though they can't provide a reason. The researchers dubbed this experience *moral dumbfounding*:

> Participants often stated immediately and emphatically that the action was wrong, and then began searching for plausible reasons. Participants frequently tried to introduce an element of harm, for example by stating that eating dog meat would make a person sick, or by stating that a person would feel guilty after voluntarily using her flag as a rag. When the interviewer repeated the facts of the story (e.g., that the dog was thoroughly cooked so no germs were present), participants would often drop one argument and begin searching for another. It appeared that judgment and justification were two separate processes; the judgment came first, and then justification (Haidt, Bjorklund, & Murphy, 2000, p. 3).

You saw a similar mechanism with congruence: *Hmm, something about this stimulus feels right. It must be [insert positive reason].*

Morality works the same way: Eating a pet feels wrong, so we confabulate reasons why it must be morally wrong. Emotions come first; *then* justification.

What kind of emotions? You might have noticed that morality has a strong connotation with feelings of disgust:

> Adults use the terms revolting, gross, and disgusting to describe entities and actions, such as feces, rotten food, and sex with corpses, which elicit a certain visceral response. But adults also apply such expressions to

certain sociomoral transgressions, such as cheating on one's spouse or stealing from the poor (Danovitch & Bloom, 2009, p. 107).

Over time, you scaffolded MORALITY onto DISGUST. Here's the historical lineage of that scaffolding (adapted from Curtis, 2007).

- Body excretions (e.g., poop, vomit)
- Spoiled food (e.g., meat)
- Certain creatures (e.g., pigs, rats)
- Certain people (e.g., lower class)
- Social and moral violations (e.g., cheating on spouse)

Today, your concept of morality is intertwined with disgust: Activating one concept will activate the other.

For example, people crinkle their nose when they feel physically disgusted (which blocks harmful input). Turns out, people also crinkle their nose when they feel morally disgusted. In one study, researchers evoked various types of disgust (e.g., bad taste, disgusting visuals, unfair treatment). All variations, including moral disgust, activated the levator labii muscle that crinkles the nose (Chapman, Kim, Susskind, & Anderson, 2009).

Even worse, this mechanism perpetuates unjust treatment toward certain groups of people. Many societies, even today, regard homosexuality as morally wrong. Physical disgust is the root cause of this prejudice: If people feel disgusted when they simulate their own homosexual tendencies, they misattribute this response to a moral prescription: *Hmm, this behavior seems disgusting. It must be morally disgusting.* When researchers dispersed a nasty smell, participants showed worse treatment toward gay men, yet they remained unchanged toward African Americans, the elderly, and other stigmatized groups (Inbar, Pizarro, & Bloom, 2012).

Or evaluate these scenarios:

- A student steals a library book that other students need to pass an exam.
- A student asks the teacher a question in class without raising her hand first.

People who become disgusted more easily perceive more wrongdoing in those scenarios (Chapman & Anderson, 2014).

Finally, consider politics. Some conservatives oppose gay marriage because that behavior seems wrong. If conservatives feel more abstract disgust, wouldn't they feel more physical disgust? Indeed, that's the case: Conservatives are more sensitive to physical disgust compared to liberals (Inbar, Pizarro, & Bloom, 2009).

Hopefully you can see the broader implications. For example, I wish I liked avocados because they're so healthy, but yuck . . . I find them gross. Since I find them disgusting, does that mean avocados are bad? Should other people stop eating avocados? Those questions seem absurd, yet—right now—society is using that same logic to structure moral codes. And that needs to change. We need to disentangle irrelevant emotions from laws and moral guidelines.

DISGUST might be a flawed paradigm for morality, but another framework—BALANCE—is more fitting. You can spot this sensory trait in the idea of fairness:

> If you can sit at the front of the bus (=an action with a specific weight), then I ought to be able to sit at the front of the bus (i.e., to balance your action with mine), insofar as we are considered equals . . . It would be an injustice (an imbalance) if you were granted a right (with a certain legal "weight") while I am denied that right, when there is no relevant difference between us (Johnson, 1987, p. 95).

That idea seems simple, but it has profound consequences in a concept that I call the *equity scale*.

The Equity Scale

Pop quiz. Which scenario would make you more likely to buy ice cream?

- You bring reusable bags
- You get insulted by a fellow customer
- Your legs get tired from walking

Give up? All scenarios can influence you to buy ice cream because of BALANCE.

Across my research, I started noticing a hidden theme of behavior: Everybody perceives a global sense of justice.

Perform a good deed? You expect to be rewarded.

Perform a bad deed? You expect to be punished.

It sounds basic, but once you understand the pieces, you'll see how our desire for balance can drastically influence our decisions.

I categorized all behaviors into four types. See Figure 16D.

- **Obligation.** Behaviors you "should" do (e.g., donate money).
- **Misdeed.** Behaviors you "shouldn't" do (e.g., punch someone)
- **Mishap.** Negative occurrences (e.g., you get punched).
- **Enrichment.** Positive occurrences (e.g., you win money).

Think of the equity scale as a balancing act: Any behavior on one side will require a behavior on the opposing side.

Many behaviors naturally counterbalance each other. An obligation (e.g., studying) might give you enrichment (e.g., good grade).

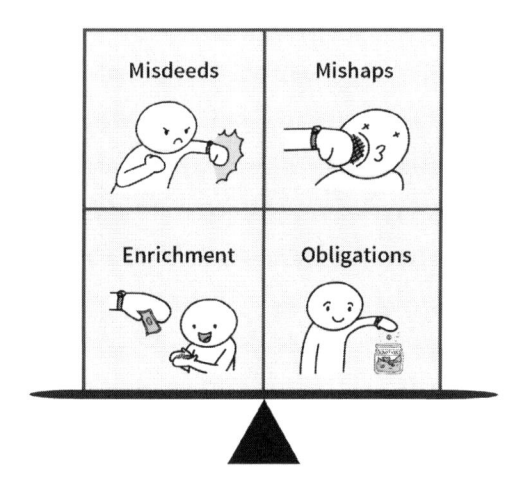

Figure 16D. The Equity Scale

On the other hand, a misdeed (e.g., cheating) might cause a mishap (e.g., expulsion). In both cases, your equity scale becomes balanced.

A problem occurs when you tilt the scale without a reciprocal adjustment. In this case, your equity scale becomes unbalanced.

Suppose that you fell down the stairs. If this mishap was accompanied by a misdeed (e.g., negligent texting), then nothing changes. However, if the mishap wasn't your fault—perhaps the floor was wet—then you feel entitled to enrichment (e.g., lawsuit) to balance your scale. The equity scale is the underlying framework behind law and justice.

Equity scales seem like common sense right now, but they have two powerful implications: immunity and absolution. For example, people described themselves with word groups:

- **Positive:** caring, generous, fair
- **Neutral:** book, keys, house
- **Negative:** disloyal, greedy, mean

Later, when asked for a donation, people donated the lowest amount of money if they described themselves with positive words. They donated the highest amount with negative words (Sachdeva, Iliev, & Medin, 2009).

- **Positive:** $1.07
- **Neutral:** $2.71
- **Negative:** $5.30

Seems contradictory, right? Wouldn't the positive group donate more money if they were reminded of their good nature? Turns out, no. Those people were reminded of past obligations. Their equity scale tilted toward the right, so they felt immune to perform a misdeed (e.g., low donation). People who described themselves with negative traits needed to perform an obligation (e.g., large donation) to regain balance in their equity scale.

The takeaway: Any behavior on one side of the equity scale—left or right—will require an adjustment on the opposing side.

Let's dive deeper into immunity and absolution.

Immunity

Behaviors on the right side of the equity scale can give you immunity to perform behaviors on the left. Perform a good deed? You gain immunity to perform bad deeds. For example, people who imagined volunteering were more likely to spend $50 on designer jeans, rather than a vacuum (Khan & Dhar, 2006).

Remember my quiz about ice cream? You are more likely to buy ice cream at a grocery store if you bring reusable bags (Karmarkar & Bollinger, 2015). That gesture of sustainability, an obligation, gives you immunity to perform a guilty behavior, like eating ice cream.

Forgot your bags? You will search the context to find

something—*anything*—that can balance this behavior. You might realize that your legs are tired from walking across the store. *Bam.* You just acquired an excuse to buy ice cream . . . you worked hard today.

Jeans and ice cream aren't too harmful, but those misdeeds can become more egregious. In one study, people who bought environmentally friendly products were more likely to steal money from the researchers at the end of the experiment (Mazar & Zhong, 2010). While taking their earnings from an envelope, people who bought green products took more money than they actually earned. Their past obligation shielded them from this misdeed.

Those biases are perpetuating immoral behavior, like racism. In one study, white people monitored their physiological responses (e.g., heart rate, blood pressure) while viewing photos of black people. If they were told that their responses didn't show a racial bias, they sat farther away from black students in a subsequent task (Mann & Kawakami, 2012). Their nonracist behavior provided immunity for racist behaviors.

You also acquire immunity from mishaps. Imagine that you're shopping in a store, and you get insulted by another customer. You just experienced a mishap. Your equity scale is tilting toward the right, which has given you immunity for a misdeed (buying ice cream).

Absolution

Behaviors on the left side of the equity scale can compel you to perform behaviors on the right. Perform a bad deed? You need to absolve this behavior with a good deed. For example, after reflecting on a misdeed, people held their arms in ice water for a longer duration, as if to punish themselves for the wrongdoing (Bastian, Jetten, & Fasoli, 2011).

Many businesses apply this principle with charity incentives: *We'll donate 10% of profits to [insert charity]*. Charity donations are more effective for emotional products (e.g., cake) because they absolve the guilt from those purchases (Strahilevitz & Myers, 1998).

Or consider reciprocity. You feel an urge to "repay" somebody for a favor because this obligation balances the enrichment. In particular, you feel two pressures: (1) an urge to perform any obligation to balance your equity scale and (2) an urge to enrich the original giver to balance their equity scale. Past research has overlooked these independent mechanisms, but we should distinguish them moving forward.

Moral Purity

Every decision can tilt your equity scale. But there's a problem. As I write this sentence, I'm 28 years old. I've made a lot of decisions during my 10,308 days of life. Has my brain tallied everything? Probably not. Instead, we reset our equity scales occasionally. How? By starting with a "clean" slate . . . literally.

As you learned, we scaffold MORALITY onto DISGUST. Therefore, we often describe moral behaviors in terms of cleaning:

> You WASH AWAY the sin of a DIRTY DEED so that you get a CLEAN SLATE.

You conceptualize immoral behaviors as impurities that you need to clean from your body. Even body parts can be contaminated. Researchers asked participants to lie to somebody. Participants who lied through email preferred hand sanitizer, whereas people who lied through voicemail preferred mouthwash (Lee & Schwarz, 2010a).

Therefore, immorality activates a need for cleansing: People who wrote about an unethical behavior preferred cleaning products (e.g., soap, toothpaste, detergent), and they completed word fragments (e.g., W_ _H; SH_ _ER; S_ _P) with cleaning words (e.g., WASH, SHOWER, SOAP; Zhong & Liljenquist, 2006).

Physical cleansing erases the moral impurity, leaving you with a "clean slate." It also erases *any* recent history, such as your equity scale. Researchers asked people to solve 25 anagrams, even though many anagrams unsolvable. During a short break, people who washed their hands believed they would receive a higher score when they resumed this task. Why? Cleansing erased the misdeed of their past failure (Kaspar, 2013b). If you bring reusable bags to a store, you might be tempted to buy ice cream until you find a hand sanitizing station.

However, there's nothing magical about water or cleaning. Cleansing merely activates a mindset. While cleaning, you are focused on separating impurities from a surface—and you extend this feeling of purity to your identity, as if you, too, are starting with a fresh foundation.

Think of PURITY as WHITE: Anything on a white background becomes immediately noticeable:

> A white object . . . is universally understood to be something that can be stained easily and that must remain unblemished to stay pure (Sherman & Clore, 2009, p. 1).

You can extend this sensory concept into personal identities. Perhaps this resemblance can explain a cultural tradition. Why do many cultures dress brides in white? Why not grooms or males?

First, you should recognize that males and females have different skin tones. Males have darker skin to maintain higher folate for sperm production, while females have lighter skin to synthesize vitamin D during pregnancy (Jablonski, 2004; Jablonski & Chaplin,

2000). You associate these colors with MALE and FEMALE: People were quicker to identify male names in dark fonts, and they were quicker to identify female names in white fonts (Semin & Palma, 2014).

This mechanism seems innocent, even cutesy, but it causes a harmful bias. If a man sleeps with many women, society praises him; yet, if a woman sleeps with many men, society condemns her. In some countries, women—even if they're raped—are ostracized or killed. Perhaps these impurities can easily be seen in the WHITE underpinning of females, yet these same impurities are hidden in the DARK underpinning of males. Disentangling these mechanisms can help us find the right solutions to these problems.

Summary

You feel empathy for people because of sensory grouping (e.g., proximity, similarity, containment). You group yourself with other people, and you believe that you are a single unit. You care for these people because you *are* them. You can increase empathy by strengthening any principle of grouping, such as visual or conceptual similarities.

You also build morality onto physical disgust. Some actions feel physically disgusting, and you confabulate reasons why they are morally disgusting.

Finally, you conceptualize fairness through sensory balance. Good actions give you immunity for bad actions, whereas bad actions compel you to perform good actions for absolution.

Other examples:

> ▶ **Dehumanizing Language.** Dehumanizing remarks (e.g., Jews are rats or African Americans are apes) are dangerous because they remove a humanlike essence, inhibiting your ability to

group and empathize. Unfortunately, this language is still prevalent in modern society. It's easier to sentence "the defendant" to death. It's harder to sentence "Kevin" to death.

- **Empathy for Animals.** People were more likely to adopt dogs who were "good listeners" (vs. "good at listening to commands;" Butterfield, Hill, & Lord, 2012). Ironically, Microsoft Word marked the previous sentence as grammatically incorrect. While referring to a dog, I was prompted to use "that" (instead of "who") because dogs aren't people. Those grammar rules are hurting animals because they remove a humanlike essence.
- **Humanlike Objects.** People treat objects better if these objects are endowed with humanlike traits. Zipcar was smart. They gave their rental cars a quirky name, like Chase, so that people would treat them better (Levine, 2009).
- **Racial Licensing.** Participants circled the US presidential candidate—Barack Obama or John McCain—for whom they would vote. People who circled Obama became more likely to hire a white person in a task (Effron, Cameron, & Monin, 2009). In a follow-up study, researchers asked people to circle the younger candidate: John McCain or Barack Obama. Despite the same action (i.e., circling Obama), people hired a black candidate because they didn't have immunity.
- **Balance vs. Consistency.** Earlier in the book, I mentioned that you possess behavioral momentum: You strive for consistency. Doesn't this contradict balancing? The full explanation is beyond the scope of this book, but the distinction involves the way you group events. If you group events together, your behavior will be consistent; if you group events separately, your behavior will be balanced. In the original study with yard signs, households might have been *less* likely to comply with the second request if they conceptualized this new request as a separate entity, perhaps due to a new researcher or location.

17

Religion

If God did not exist, it would be necessary to invent him.
—Voltaire

I WAS RAISED CATHOLIC, and I followed the religious gamut: studied the Bible, prayed to God, and attended Catholic schools across my life. I believed in God, but I never really questioned it. Everyone around me was religious, so why disrupt the status quo?

Well, that changed in college. Perhaps it was my blossoming appreciation for truth and knowledge, as well as my growing desire to understand the beauty and complexity of our universe. Or perhaps it was the heavy drinking. Either way, I started questioning my beliefs because I wanted to find the truth. Not a comforting belief for my death. Not the same beliefs of my peers. Not a logical belief of atheism. But the *truth*.

I scoured research from all perspectives, lined up the arguments, and derived my conclusion. To me, the answer was abundantly clear: God didn't create humans; humans created God.

This chapter explains why humans were predisposed to create religion, but it doesn't contradict the existence of God. If you believe in God, you could argue that God instilled these mechanisms. I disagree with that argument, but my rebuttal—as well as the reasons behind my atheistic conclusion—are beyond the scope of this chapter.

I'll be using Christianity as the exemplar religion, but the concepts will apply to any religion.

Three Goals of This Chapter

This chapter isn't trying to eradicate religion from society, but it's striving toward three changes:

- **We should be able to discuss (and criticize) religion.** Religion is sacred to many people, and we should respect—and, in some cases, admire—those beliefs. People should have the freedom to pursue any religion, as long as those beliefs don't impede human rights or societal progress. The problem? Some beliefs *are* impeding human rights and societal progress. Yet we've ignored these detriments because of the protective shield of religion. That needs to change. We need to remove that shield. We need to discuss (and criticize) religious beliefs when they become detrimental to society.
- **We shouldn't derive science from religion.** We should support freedom of belief, but we shouldn't slow societal progress because of an obligation of toleration. Some people truly believe that earth is flat. Should we add that belief into school curriculums? No. We should recognize when certain ideas (e.g., creationism) contradict the nature of science, and we should remove them from curriculums.
- **We shouldn't derive morality from religion.** Many people assume that religion dictates morality. But, if anything, morality has dictated religion. Over time, religious leaders have been forced to broaden interpretations of holy texts to accommodate our growing understanding of morality (e.g., slavery, gay rights, premarital sex). Religious beliefs are hindering the progression of our moral compass. Instead of waiting for

religious beliefs to catch up, we should sever the ties between morality and religion.

Human Origins of Religion

Many religious concepts are grounded in earthly metaphors. You've seen a few examples, such as location (e.g., HEAVEN is UP; HELL is DOWN).

When you keep digging, you spot other earthly concepts. For example, most Christians believe a core doctrine: In order to be saved, you need to believe in God. If you don't accept Christ, then you are going to hell. Christians are following their OBJECT primitive by creating two discrete buckets:

▶ Yes, I believe in God.
▶ No, I don't believe in God.

However, belief isn't dichotomous. Belief is a spectrum. People can possess *some* belief in God, which complicates the doctrine. For example, what percentage of belief is acceptable? Do you need a little belief, like 25%? Or do you need to pass a threshold, say 51%? Or maybe you need a full and resolute belief, like 100%? Could you miss the cutoff by a few percentage points? That'd be unlucky, huh? Not to mention, *when* is our belief judged? Our beliefs frequently change based on many factors (e.g., context, time, day, mood). Heck, your beliefs are different *within* your brain. Researchers asked split-brain patients whether they believed in God: Their right hemisphere indicated "no" while their left hemisphere indicated "yes" (Ramachandran, 2006). Would the left hemisphere be saved? Again, when would God judge these fluctuating beliefs? Are we judged at our death when we're senile with dementia? What if we convert *because* of dementia? Would that still count? Or perhaps the judgment is averaged across our lifespan. If we haven't believed by 60 years old, should we throw in the towel?

Those questions seem nitpicky—and they are—but let's expand the implications toward something important.

Suppose that religious beliefs are truly dichotomous. Even with these two outcomes—belief vs. disbelief—we still can't control those beliefs. Don't look now, but I'll give you a million dollars if you believe that a pink unicorn is galloping behind you. A million bucks! With that large incentive, surely you can adopt that belief, right? You *want* to believe—and you might *claim* to believe—but would you *actually* believe? Probably not. Large incentives (e.g., one million dollars, eternal salvation) provide *motivation* to believe, but they don't provide the *ability* to believe. Based on the doctrines of Christianity, people could seek God for an entire lifetime: go to church every Sunday, read the Bible every day, and live a morally righteous life. Yet, without adopting a genuine belief—an ability that lies outside of their cognitive control—those people are "immoral" (and will suffer eternal torture).

Would an all-loving God create conditions for inescapable failure? Or are humans sculpting religion from earthly concepts?

Let's start dissecting evidence for the latter.

Origin of Anthropomorphic Gods

Religious texts often depict God with human traits, such as the "right hand" of God. But if humans developed their morphology (e.g., hands) from evolutionary conditions on earth, why would God—a deity that doesn't live on earth—possess traits that earth sculpted? He wouldn't.

You could argue that the language is metaphorical, but this section will explain how and why we created God in our image. It involves the two motivations that cause us to anthropomorphize: sociality motivation and effectance motivation.

Sociality Motivation

If you desire a social connection, you attribute human traits to inanimate objects:

> ... those who lack social connection with other humans may try to compensate by creating a sense of human connection with nonhuman agents (Epley, Akalis, Waytz, & Cacioppo, 2008, p. 114).

For example, lonely people buy more brands with anthropomorphic spokespeople because they are craving any social connection (Chen, Wan, & Levy, 2017).

Or consider these gadgets:

- **Clocky.** An alarm clock that "runs away" so that you chase it.
- **CleverCharger.** A battery charger that prevents overheating.
- **Pillow Mate.** A torso-shaped pillow that gives hugs.

People attributed more human traits (e.g., free will) to those gadgets if they felt lonely (Epley et al., 2008). They solved their loneliness by confabulating a social connection with them.

You find the same effect in religion. God is an easily accessible social connection for anyone who craves sociality. Therefore, you are more likely to seek God if you are:

- Lonely (Kirkpatrick, Shillito, & Kellas, 1999)
- Single (Granqvist & Hagekull, 2000)
- Widowed (McIntosh, Silver, & Wortman, 1993)

The reverse happens, too: If you feel rejected by God, you seek humans. When researchers asked participants to reflect on personal traits that God would dislike, they reported closer relationships with their romantic partners (Laurin, Schumann, & Holmes, 2014).

Effectance Motivation

Ambiguity makes you feel anxious, and you overcome this anxiety by seeking control, knowledge, or security. How? You transform something unknown into something familiar.

To solve the unknown mysteries of the world, such as human existence, our ancestors generated an anthropomorphic hypothesis: God. How can we tell? Let's look more closely at two components of effectance motivation: control and security.

Control. You live in a chaotic world. Negative events, like earthquakes, can make you feel weak and helpless, as if you can't control the events around you.

Belief in God, an all-powerful deity, imbues the chaotic world with purpose and order—which can feel comforting. Before elections, a time period filled with uncertainty, people indicate stronger beliefs in God because they seek control and purpose (Kay, Shepherd, Blatz, Chua, & Galinsky, 2010). In fact, researchers analyzed historical periods with greater threat (e.g., social, economic, political). During uncertain times, humans gravitate toward churches that are more authoritarian because they prefer a stricter belief system (McCann, 1999).

If you can't find control from religion, you search for other sources of control. After reading an article that doubted the existence of God, people showed stronger support for the government (Kay et al., 2010).

If both sources—God and government—are unavailable, you perceive more control in yourself. In one study, people pressed a button to generate a green circle, yet the circle only appeared 75% of the time. If they watched a video that depicted the government as incapable of restoring order, they believed they had more control in generating the circle (Kay, Gaucher, Napier, Callan, & Laurin, 2008).

You seek control when you feel helpless, which can explain a perplexing effect. After devastating events, such as a deadly earthquake, people lose faith in God across the world except in one place: The area that was afflicted. In those damaged regions, people become *more* religious (Sibley & Bulbulia, 2012). Same with other devastating events: Parents who lose a child become *more* religious (McIntosh et al., 1993).

Intuitively, these events should weaken faith. Why would an all-loving God commit such a terrible act? Yet these events *increase* faith because people are searching for meaning and purpose. Why did it happen? Belief in God, an all-powerful deity, can make people feel like the event has a hidden purpose (Schjoedt & Bulbulia, 2011). Believing that your child died for a higher purpose is more comforting than believing the alternative.

Security. Religion imbues the chaotic world with meaning, but it also offers safety and security. Among 137 countries, people who live in more unstable conditions (e.g., health, income) are more likely to be religious (Barber, 2011; Rees, 2009).

Religious beliefs are particularly appealing because they offer immortality after our death:

> . . . religious beliefs are particularly well suited to mitigate death anxiety because they are all encompassing, rely on concepts that are not easily disconfirmed, and promise literal immortality (Vail et al., 2010 p. 84).

Participants who wrote about death were more likely to indicate a stronger belief in God and Shamanic spirits (Norenzayan & Hansen,

2006). Inevitable death is scary, so we calm our anxiety by pursuing a belief in immortality.

How Gods Became Moral

We conceptualize gods with human qualities. But why *moral* gods?

That question might sound weird. Isn't *that* religion? Aren't all gods moral? Actually, no. Back in the day, gods never cared about human affairs, let alone moral affairs. We merged morality with religion about 12,000 years ago (Norenzayan, 2014). So, why the transition?

Moral gods offered a better way to govern society. Until this point in history, humans cooperated with each other through reciprocal altruism: You help me, and I'll help you (Trivers, 1971). That style of governance was successful in small societies where everybody knew each other, but it failed in larger societies because of the interactions with anonymous strangers. Why help somebody that you'll never see again? There was no incentive.

Societies needed a better a system of governance that could (a) set proper rules, while (b) monitoring and punishing rule breakers even in private. Eventually, societies stumbled upon the right formula: Moral gods that were omniscient (could see everything) and omnipotent (could punish accordingly).

You can see the remnants of that history in the modern world. Today, you're more likely to find moral gods in larger societies:

> . . . although all known societies have gods and spirits, there is a cultural gradient in the degree to which they are (1) omniscient, (2) interventionist and (3) morally concerned . . . As group size increases, the odds increase of the existence of one or several Big Gods—omniscient,

all-powerful, morally concerned deities who directly regulate moral behavior and dole out punishments and rewards (Norenzayan, 2014, pp. 374–375).

Moral gods are especially prevalent in areas with hard-to-govern resources, like water supplies (Snarey, 1996; Johnson, 2005). In foraging societies that require little cooperation and oversight, moral gods are virtually nonexistent (Peoples & Marlowe, 2012).

Let's briefly walk through the step-by-step history of moral gods (see Kiper & Meier, 2015; Norenzayan, 2013).

1. People created anthropomorphic gods

Again, humans live in a chaotic world. We reduce our anxiety by conceptualizing a deity that oversees everything, and we perceive this deity with human traits because of our instinctive need for social connection.

2. People earned more trust if they believed in moral gods

Our ancestors didn't have Facebook or LinkedIn, but they still needed a way to judge the trustworthiness of anonymous strangers. So, how did they do it?

Well, if people believed in moral gods—especially gods that could intervene and punish their wrongdoing—wouldn't *you* trust them?

These beliefs became a social glue. Even if people believed in different gods, they could still cooperate with each other because they were held accountable to an external source.

One problem, however, was that people could fake their belief: *Yeah, sure. I believe in a moral god.* Religion solved this problem by enacting strange rituals and customs:

> The extravagance of some religious rituals has long puzzled evolutionary scientists. These performances demand sacrifices of time, effort, and resources. They include rites of terror, various restrictions on behavior (sex, poverty vows), painful initiations (tattooing, walking on hot stones), diet (fasts and food taboos), and lifestyle restrictions (strict marriage rules, dress codes; Norenzayan, 2016, p. 12).

These rituals identified fakers. You could tell who was genuinely adhering to religious beliefs, and you could trust those people because of their external accountability.

3. Societies with moral gods could expand to farther geographies

Moral gods helped societies expand geographically because of two byproducts: monitoring and punishment.

First, people act morally if they believe that other people are watching (Piazza, Bering, & Ingram, 2011). Big and powerful gods extracted moral behavior in society (even in private) because it felt like you were being watched:

> The omniscience of these agents extend one's vulnerability to "being caught" to all times and all places. Some gods can even read people's thoughts . . . The consequence is that "hidden defection," which was still a viable individual strategy in groups with indirect reciprocity, is markedly reduced (Shariff, Norenzayan, & Henrich, 2010, pp. 12–13).

The second component is punishment. In game theory, punishment is required for maximal cooperation (Johnson & Krüger,

2004). Even today, people behave morally if they believe in a deity that will punish their wrongdoings (Purzycki et al., 2016).

Both concepts—monitoring and punishment—offered a better system of governance than reciprocal altruism. Individuals could finally cooperate with anonymous strangers, allowing society to expand into farther geographies.

4. Society replicated the purpose of moral gods via laws

Godly governance was effective in beginning stages because it was cheaper than human governance. Eventually, though, we started replicating this governance through laws, courts, and other institutions. Modern societies have solved the problems that religion was designed to solve—which begs an important, yet difficult, question: Do we still need religion? I'll give my thoughts at the end of this chapter.

To recap, humans created anthropomorphic gods to solve fundamental needs (e.g., sociality, control, security). Societies with particular gods—moral, omniscient, omnipotent—could outsource their governance to these gods, which allowed inhabitants to expand to farther geographies. People could finally cooperate with anonymous strangers because everybody was held accountable for their actions.

Implications of Moral and Humanlike Gods

You live on a planet with human beings. Even though God doesn't live on this planet, you conceptualize this deity with humanlike traits, including a humanlike mind:

> God is perceived to have more or less of certain abilities, but God is not perceived to have an entirely unique sort

of mind with capacities that are unheard of in human minds. For example, it appears nonsensical to debate whether God's mind can fly, because that is not the kind of thing that a (human) mind does (Heiphetz, Lane, Waytz, & Young, 2016, p. 13).

God has a human mind. And thus, your ability to mentalize—the act of deciphering people's thoughts—will apply to God. Evidence shows that women, who are better mentalizers, are more likely to believe in God (Roth, & Kroll, 2007). People with autism can't mentalize, so they naturally have weaker beliefs in God (Norenzayan, Gervais, & Trzesniewski, 2012).

However, the most striking examples come from your own mind. Remember projective simulation? You decipher people's thoughts by inserting yourself into their mind, and you adjust (insufficiently) from your own beliefs.

You exhibit that same bias with God: You insert your own beliefs into the mind of God. In one study, researchers asked people to rate their beliefs on various topics (e.g., abortion, death penalty, same-sex marriage). Then, they rated the beliefs for God and various people (e.g., Bill Gates, Barry Bonds, George Bush, average American). Sure enough, God's beliefs were eerily similar to their own beliefs (Epley, Converse, Delbosc, Monteleone, & Cacioppo, 2009).

And it gets worse. If you truly insert your mind into God, then changing your beliefs should change God's beliefs (but not the beliefs of other people). In another study, participants read arguments that supported or opposed affirmative action. Lo and behold, participants believed that God—but not other people—shared their newly adjusted belief (Epley et al., 2009).

Still not convinced?

Researchers used fMRI to study the neural patterns of people while they reflected on the beliefs of various entities (e.g., themselves, God, other people). When they reflected on God's beliefs, those neural patterns matched the neural patterns of their own beliefs (Epley et al., 2009).

You aren't created in God's image. God is created in *your* image.

Science and Religion

Many Christians reject evolution in favor of creationism. And this blatant rejection of science creates a dangerous ideology for society.

But first, let's peel back the layers: Why is creationism intuitively appealing? You can blame *teleology*.

Today, our world is flooded with tools: knives for cutting, chairs for sitting, books for reading, and the list goes on. By living in a world surrounded by tools—tools with a clear purpose—children learn causal reasoning through teleological explanations, as if objects were "made for something." Those explanations are accurate for artificial objects (e.g., chairs for sitting), but they misrepresent natural objects (e.g., mountains don't exist for climbing; the sun doesn't exist for warmth; Kelemen & Rosset, 2009; DiYanni & Kelemen, 2005).

Creationism is the quintessential teleological explanation: God created the earth for humans. In addition, many religious people ask: Why are we here? They believe that each human has a purpose in life, as if we are tools designed for something.

Eventually, children learn to think critically, but they build these new skills onto an existing foundation of teleology. Ultimately, teleology remains the foundation of causal reasoning.

> . . . while the acquisition of scientifically warranted causal explanations might suppress teleological ideas, it does not replace them (Kelemen & Rosset, 2009, p. 2).

Critical thinking is built onto a teleological foundation. Therefore, if you can't access these advanced skills, you rely on teleology.

For example, why does rain exist? You can answer this question in two ways:

- ▶ **Mechanistic:** Water condenses in clouds to form droplets
- ▶ **Teleology:** Plants and animals need water for drinking

Patients with Alzheimer's disease preferred the teleological explanation (Lombrozo, Kelemen, & Zaitchik, 2007). So, too, did healthy adults while answering quickly (Kelemen & Rosset, 2009). Dementia and quickness inhibited their advanced thinking, so they used their default mode of reasoning (teleology).

Similar effects happen with religion. Humans possess two modes of reasoning: System 1 is fast, while System 2 is slow and effortful (Kahneman, 2011). Religious people, who prefer a default mode of teleology, rely on System 1 more often.

Answer this question:

A bat and ball cost $1.10 in total. The bat costs $1.00 more than the ball. How much does the ball cost?

Got your answer?

Religious people are more likely to get this question wrong (Shenhav, Rand, & Greene, 2012). They choose the seemingly obvious answer of $0.10 (instead of the correct answer of $0.05) because they rely on System 1. Perhaps this insight could explain why intelligent people, who rely more heavily on System 2, are less religious (Zuckerman, Silberman, & Hall, 2013).

But who cares, right? Religious people aren't hurting anybody. Who cares whether or not they believe in science?

Unfortunately, a blatant dismissal of science fosters a dangerous ideology in society. Many religious people don't believe in climate change (Hope & Jones, 2014). If anything happens, God will restore creation. This philosophy is dangerous, and we need to separate religious views from science.

We need to combat scientific misunderstanding through proper education. In the US and countries with Islamic fundamentalism, religious extremists insert creationism into science classrooms. But, much like a flat earth, some "possibilities" should be excluded. We should teach creationism, but we should teach it in the proper classroom (e.g., history, religion). If we loosen scientific criteria, then we cultivate societies that feel justified in rejecting scientific evidence, fueling dangerous ideologies.

Morality and Religion

Are religious people more ethical than nonreligious people? Many researchers disagree on this answer. I'll give you arguments on both sides, and then I'll merge everything into my assessment.

Religious beliefs *can* increase morality, but it depends on the saliency and content of these beliefs.

First, religious concepts need to be activated and salient. Religious people were more likely to donate, but only on days of worship (when their concept of morality was activated; Malhotra, 2008).

Second, it also depends on the content of religious beliefs, such as the portrayal of God. In one study, people read Bible passages that described God as forgiving or punishing. People who reflected on a forgiving God were more likely to steal money from the researchers (DeBono, Shariff, Poole, & Muraven, 2017). I assume the researchers were less forgiving.

Despite those benefits, however, religion also promotes outdated beliefs that hurt other people. One problem is prejudice. Subliminally exposing people to religious concepts increased their likelihood of derogating outgroups (e.g., atheists, Muslims, gay men,

African Americans; Johnson, Rowatt, & LaBouff, 2012; Johnson, Rowatt, & LaBouff, 2010).

Prejudice toward gay people is especially strong. In one study, religious people read essays by fellow students: one essay praised technological progress, while another essay—written by a gay student—praised the growing acceptance of gay rights. Later, in an unrelated task, religious people forced the gay author to eat more hot sauce, as if to punish a wrongdoing (Blogowska, Lambert, & Saroglou, 2013).

Hot sauce might be harmless, but worse effects occur in countries with stronger beliefs (Kohut et al., 2013). Some Christians try to "fix" gay people through conversion therapy, which is ineffective and psychologically damaging (Jenkins & Johnston, 2004).

Clearly, religious people can be immoral. But what about atheists? Surely, they must be worse? According to a national survey, atheists garnered more resentment than all other groups:

> ... atheists are less likely to be accepted, publicly and privately than any others from a long list of ethnic, religious, and other minority groups (Edgell, Gerteis, & Hartmann, 2006, p. 211).

Yet, when you look at the data, atheists are actually more ethical than religious people (Meier, Fetterman, Robinson, & Lappas, 2015).

> ... atheists and secular people actually possess a stronger or more ethical sense of social justice than their religious peers ... when it comes to such issues as the governmental use of torture or the death penalty, we see that atheists and secular people are far more merciful and humane. When it comes to protecting the environment, women's rights, and gay rights, the non-religious again

distinguish themselves as being the most supportive. And as stated earlier, atheists and secular people are also the least likely to harbor ethnocentric, racist, or nationalistic attitudes (Zuckerman, 2009, p. 954).

Here's the point: In addition to separating science and religion, we should also separate morality and religion. Instead of twisting the meaning of religious texts to accommodate our evolving moral compass, we need to sever the ties. We need to recognize that many religious doctrines are inherently immoral, and we need to remove these outdated philosophies. We need to cultivate societies based on true morality.

Summary

Across evolution, human morphology has advanced so much that we have vestigial organs: Some adaptations (e.g., appendix, wisdom teeth, male nipples) are no longer useful.

I believe that religion has become a vestigial organ.

Societies, throughout history, created many religions that weren't adaptive. They didn't help us survive and thrive. Eventually, though, we found the right formula: moral gods who were omniscient and omnipotent. Societies could expand into farther geographies because humans could now cooperate with anonymous strangers who were held accountable.

Over time, we began replicating those benefits in other ways (e.g., laws, police, government). Today, religion has lost the key purpose for which it was designed, which leads to a difficult question: What should we do?

On one hand, societies might fare better without religion. Across the world, religiosity is correlated with higher murder rates

(Jensen, 2006). Even US states with more religion (e.g., Louisiana, Alabama) have higher murder rates, while states with less religion (e.g., Vermont, Oregon) have lower murder rates (Zuckerman, 2009).

On the other hand, religion has potential. Despite my atheistic beliefs, I see the value in religion. I'm hoping that society can replicate the benefits of religion—a strong community, sense of purpose, safe haven, comforting beliefs in death, and more—while removing the inaccurate (and often judgmental) beliefs that follow.

And I think we should start with these goals:

- ▶ We should be able to discuss (and criticize) religion.
- ▶ We shouldn't derive science from religion.
- ▶ We shouldn't derive morality from religion.

These goals will help disentangle the detriments. Afterward, we can start fostering societies that treat all people with dignity and respect—not because of rewards or punishment, but because of genuine compassion.

18

Literature

COMPARE these two sentences:

- ▶ John painted houses last summer.
- ▶ John was painting houses last summer.

Same meaning, right? Not quite.

In this chapter, you'll learn why subtle words can alter the meaning of sentences. You'll learn what's happening inside your brain as you read this sentence. And this one too. Not only will you learn why certain books can sweep you away, but you'll also discover techniques to improve the clarity of your writing.

Simulations of Words

While reading sentences, you construct meaning by generating simulations of each word. Reading the word "bird" activates a canonical image, spatial location, chirping sound, and other common traits of birds.

One simple word activates a barrage of imagery. So, imagine a more descriptive sentence: *The bird plucked a fish from the ocean.*

Holy moly... that sentence may have seemed innocent beforehand, but hopefully you see the fireworks of simulations that are occurring in your brain. You are activating past experiences with all concepts (i.e., bird, fish, ocean), along with a motor action of plucking. As you'll see next, subtle words can influence these mental images.

Verb Aspect

Verb aspect determines whether an action is completed or ongoing. Fill in these blanks:

- When John walked to school, _____ .
- When John was walking to school, _____ .

Those sentences seem identical, yet the second sentence is more vivid because you construct a mental image in which John is currently walking to school. Participants filled in those blanks with more actions (Matlock, 2011).

Remember these sentences?

- John painted houses last summer.
- John was painting houses last summer.

In the second sentence, you believe that John painted more houses (and spent more time painting) because it depicts the ongoing nature of the activity (see Figure 18A; Madden & Zwaan, 2003; Matlock, 2011).

Minor words can make a dramatic impact. During legal trials, an attorney could communicate two depictions of a suspect's behavior:

- Keith pointed his gun.
- Keith was pointing his gun.

|Imperfective|Perfective|

Figure 18A

Juries are more likely to convict Keith in the second version because the mental imagery is more vivid (Hart & Albarracín, 2011).

It gets worse with conditionals.

Conditionals

If you win the lottery, what would you do with the money? Keep this in mind.

Earlier, I explained that simulations seem real. When people simulate eating M&Ms, they eat fewer M&Ms from a bowl because they satisfied their real-world desire (Morewedge et al., 2010).

This trickery occurs with if-then statements. In order to simulate your spending behavior with the lottery question, you needed to imagine winning the lottery (see Figure 18B). This example was purely hypothetical, yet—right now—you believe that you're more likely to win the lottery:

> . . . if–then statements trigger a mental simulation process in which people suppose the antecedent (if statement) to be true and evaluate the consequent (then statement) in that context . . . evaluating a conditional will heighten belief in its antecedent more than in its consequent (Hadjichristidis et al., 2007, p. 2052).

Figure 18B. My simulation of winning the lottery.

Attorneys could sway a verdict by asking juries to imagine a hypothetical supposition of guilt:

If the accused is guilty, _____ .

Juries construct a mental image in which the defendant is guilty—and this imagery, by itself, biases their verdict toward guilty.

We claim that suspects are innocent until proven guilty, yet the verdict of "Not Guilty" contradicts this right. Juries understand this statement by simulating guilt (and the subsequent removal of it). We need a label that circumvents this imagery of guilt. We need a term that constructs an image of innocence.

Syntax

The same words can instill different images based on their order and sequence. Compare these sentences:

- I taught Harry Greek.
- I taught Greek to Harry.

When "taught" and "Harry" are close together, you simulate Harry learning Greek—and you believe that he learned the lesson (Lakoff & Johnson, 1980).

When "taught" and "Greek" are close together, you simulate the content of the lesson. Harry lies outside of this primary image, so you are less likely to believe that he learned the lesson.

Or consider these questions:

- ▶ **Variation A.** What is the probability that the hottest day of next week will be Sunday?
- ▶ **Variation B.** What is the probability that Sunday will be hotter than any other day next week?

It seems like the same question, yet people estimate a higher probability in the second question (Fox & Rottenstreich, 2003). Can you spot the difference?

Each question activates a different quantity of options:

- ▶ **Variation A.** Sunday, Monday, Tuesday, Wednesday, Thursday, Friday, or Saturday
- ▶ **Variation B.** Sunday or any other day

Variation A has seven options, whereas Variation B has only two options. In both variations, you distribute the probability equally, allocating a greater portion of probability to Sunday in Variation B.

Distance

Similar objects are frequently close together:

> Imagine picking one flower in a field of various wildflowers and then picking another flower that is either right next to it or one that is 10 paces away. The closer flower

is more likely to belong to the same species as the first (i.e., the same category of flowers) and, therefore, to be more similar (Casasanto, 2008, p. 1054).

You construe this effect in writing, too. Proximity can affect the clarity and vividness of sentences. Compare these sentences:

- ▶ Mary is not happy.
- ▶ Mary is unhappy.

Mary seems less happy in the second sentence because the prefix un- is closer to happy (Lakoff & Johnson, 1980). You merge these ideas more easily. Similarly:

- ▶ Jamal found that the chair was comfortable.
- ▶ Jamal found the chair comfortable.

In the second sentence, the chair seems more comfortable because "chair" and "comfortable" are spatially closer.

Seamless Continuity

Until now, most simulations have depicted a single image:

> The bird plucked a fish from the ocean.

You understood that sentence by constructing a single image of the event. However, read this passage:

> The bird plucked a fish from the ocean. As the bird flew away, it devoured the meal.

Uh oh . . . that passage has multiple events. You need to generate multiple images to decipher the meaning. I'll refer to this idea as *dynamic simulation*: You combine multiple images to depict a series of events.

You can add new images in two ways:

- **Incremental.** You attach information to existing simulations.
- **Global.** You abandon simulations to start a new simulation.

Writers should strive for incremental changes, which are easier to process (Ohtsuka & Brewer, 1992).

Certain words and syntax can help readers integrate new information into existing simulations. I call it *seamless continuity*.

Here are some techniques.

1. Active Continuity

Active sentences are easier to process than passive sentences. Compare these simulations:

- The bird plucked a fish from the ocean.
- A fish was plucked from the ocean by the bird.

The second sentence is passive; it places the subject later in the sequence. While reading that sentence, you imagined a fish being pulled from the ocean by a mysterious entity. Once you discovered the subject—the bird—you needed to retroactively update your simulation with this new detail.

Active sentences are easier to process because you can attach each new piece to the unfolding simulation in real time.

Side note: These tips are only guidelines. Once you understand them, you can break them when appropriate. Sometimes I write passive sentences if I want to emphasize the recipient of the action (in this case, fish).

2. Descriptive Continuity

Descriptive words paint stronger images: *The majestic bird plucked an unsuspecting fish from the vast blue ocean.*

Vague words can leave gaps in your simulation:

- ▶ The journalist began the article.
- ▶ The journalist wrote the article.

The first sentence is harder to read because it creates a blurry section in your simulation (Hagoort, Baggio, & Willems, 2009).

Personally, I avoid these syntaxes too:

- ▶ There are _____ .
- ▶ This is _____ .

How do you picture those abstract beginnings? Sure, you might use sensory concepts to provide a framework, but why not choose something more concrete? This tangibility is easier to process.

Read the previous sentence again. I deliberately inserted a noun into the beginning of the sentence: "This *tangibility* is easier to process." Imagine if I excluded this noun: "This is easier to process." Your imagery would be more abstract because it doesn't have a concrete entity that depicts the action.

3. Referential Continuity

New sentences should often begin by referencing the end of a previous sentence. For example, researchers gave two types of instructions to replicate a layout:

▶ **Continuous.** The knife is in front of the pot. The pot is on the left of the glass. The glass is behind the dish.
▶ **Discontinuous.** The knife is in front of the pot. The glass is behind the dish. The pot is on the left of the glass.

Both instructions provide the same layout, yet people made more mistakes with the second version (Ehrlich & Johnson-Laird, 1982). They needed to hold separate information in working memory:

> . . . subjects try to integrate each incoming sentence into a single coherent mental model . . . those sentences which cannot be immediately integrated are represented in a propositional form (Ehrlich & Johnson-Laird, 1982, p. 296).

Here's an example in a realistic context:

▶ **Continuous.** I've been discussing referential continuity. This concept helps you simulate an unbroken chain of events.
▶ **Discontinuous.** I've been discussing referential continuity. You can simulate an unbroken chain of events because of this concept.

The first example offers seamless continuity, whereas the second example breaks that continuity.

4. Connective Continuity

Certain words can help you bind new sentences into an existing simulation. Participants scored higher on reading comprehension when researchers inserted "because" into passages:

▶ **Slower.** Most of them do not bite. Degeneration of their mouth parts enables them to feed on flower nectar only.

▶ **Faster.** Most of them do not bite because a degeneration of their mouth parts enables them to feed on flower nectar only.

Some researchers call them *connectives* or *coherence markers* (Kamalski, 2007). You encounter various types:

▶ **Additive:** and, or
▶ **Temporal:** then, next
▶ **Causal:** because, so
▶ **Adversative:** but, though

Connectives help you bind new information, so incorporate those subtle words throughout your writing (like the "so" in this sentence).

5. Expected Continuity

You've probably heard the common advice to avoid writing sentences that end with prepositions: *at, in, on, off, to, for, over, under.*

▶ **Wrong:** What time are you leaving at?
▶ **Better:** At what time are you leaving?

Grammatically, both are fine. Aesthetically, the first (well, both) need improvement. The second example might be slightly better in written prose because it has *expected continuity.*

When you reach the top of a staircase, sometimes you mistakenly believe that another step exists. If you don't encounter that step, you awkwardly stumble (and perhaps you casually repeat that subtle motion a few times so that other people don't think you're a lunatic).

Prepositions are triggering a similar effect. You typically see them *before* objects:

▶ Throw this book *at* _____ .

- ▶ He jumped *in* _____ .
- ▶ You chase him *over* _____ .

Prepositions usually come before something. Whenever you see a preposition, you prepare to simulate an upcoming object. However, if the preposition appears at the end of a sentence, there is nothing left to simulate—which feels jarring (like an extra step that doesn't manifest).

Expected continuity can also explain a frustrating part of writing. You pour your heart into a piece of writing, and it sounds wonderful. A few weeks later, you read it again. Yikes . . . what happened? Why does it sound awful?

Your writing sounds great in the initial stages because you are familiar with the message and syntax. You can predict the upcoming sequences in the simulation, and you misattribute this fluency with better writing. You need time to forget the syntax and takeaways so that you can proofread without any preconceived expectations.

6. Singular Continuity

Some sentences can have multiple interpretations:

> Mary bought her daughter a new backpack because she was having a bad day.

In that sentence, the pronoun "she" is ambiguous. Most readers will assume that Mary's daughter was having a bad day, but the mere existence of a second possibility (i.e., Mary was having a bad day) degrades the clarity of your simulation. The following passage clarifies that ambiguity:

> Mary noticed that her daughter was having a bad day, so she bought her a new backpack.

That passage has *singular continuity* because it eliminates a competing simulation.

7. Temporal Continuity

You simulate time in sentences. Recall these sentences:

> ▶ John opened the book, and an hour later, he finished it.
> ▶ John opened the book, and the next day, he finished it.

In the second example, you're more likely to erase the original simulation to create a new simulation (Radvansky, Zwaan, Federico, & Franklin, 1998). You can't remember opening the book as vividly because that event resides within an abandoned simulation.

Time can also be implied. In a fictional story, researchers described Joe who was walking from the library to dinner, and they alternated two sentences about a stick that Joe found along the way:

> ▶ He carved the stick into a small flute.
> ▶ He carved his initials right on the stick.

Participants who read the flute sentence had a weaker memory for the beginning of the story because Joe would have spent more time carving a flute. These readers created a simulation to reflect this longer duration (Rapp & Taylor, 2004).

You see the same effect with flashbacks. Consider four events in chronological order:

> E1, E2, E3, E4

Suppose that E1 appeared as a flashback later in the narrative.

> E2, E3, E1, E4

Most people assume that you remember information more easily when it occurs more recently. But that's not the case. Even though E1 occurs more recently than E2, you don't remember it more vividly. Your comprehension adheres to the chronology of events.

Don't believe me? Researchers manipulated the perceived length of E2 in that sequence (E2, E3, E1, E4). When E2 was longer, people had worse memory for E1 because this event would have occurred in a more distant past (Claus & Kelter, 2006).

People

While reading sentences, you also immerse yourself *into* people. Read this sequence:

> Mr. Ranzini was sitting outside on his front stoop.
> He had lived on this block for over 30 years.
> Next door was a local playground for the children.
> Directly across the street was the mailbox that he used.
> As usual, Mrs. Rosaldo was taking her poodle for a walk.
> Suddenly, a large truck pulled up in front of Mr. Ranzini.

Now, *quick*: Was there a mailbox across the street?

You immersed yourself into Mr. Ranzini while reading that passage. You experienced the narrative world from *his* viewpoint. You were slower to remember the mailbox because a large truck drove in front of Mr. Ranzini, which blocked *your* viewpoint. For some groups, researchers replaced that truck with a smaller object:

> Suddenly, a man on a bike rode in front of Mr. Ranzini.

Participants answered the question faster because the mailbox was still visible in their simulation (Horton & Rapp, 2003).

244   The Tangled Mind

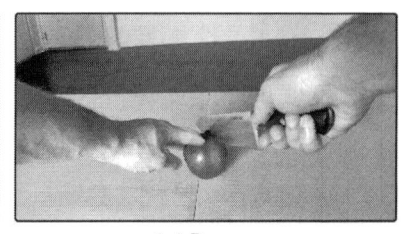

| 3rd Person | 1st Person |

Figure 18C

Oftentimes, your simulation depends on the pronouns in a sentence. For example:

- ▶ You are slicing an apple.
- ▶ He is slicing an apple.

People who read "you" pronouns constructed a mental image with a first-person perspective (see Figure 18C; Brunyé, Ditman, Mahoney, Augustyn, & Taylor, 2009).

Narrative Transportation

While reading narratives, you sometimes forget the world around you because you transport yourself into the story. Researchers describe this immersion as *narrative transportation* (Escalas, 2004).

One key factor is the availability of your self-concept. You feel weaker immersion into a story when mirrors are nearby because you can't escape your body (Kaufman & Libby, 2012).

You feel stronger immersion without your self-concept. In one study, researchers displayed this message:

> . . . we are not interested in you as a member of the college student population. We are running this study in order to assess the attitudes and perceptions of students

in general. For the purposes of today's study you represent an average student... we have assigned you an arbitrary code number for this session: SLREP51 (Kaufman & Libby, 2012, p. 5).

That message stripped away their identity, so people immersed themselves into a story more easily (Kaufman & Libby, 2012).

When you immerse yourself into characters, you *become* them. Children spent more time with a Rubik's cube after reading a story about a professor, and adults performed worse on an exam after reading a story about a "stupid soccer hooligan" (Dore, Smith, & Lillard, 2017; Appel, 2007).

Since you *become* characters, you adopt "good" traits when you immerse yourself into good people. Children who frequently read Harry Potter were kinder toward stigmatized groups (e.g., immigrants, refugees, LGBT). Similar effects occurred with adults: White people showed less racism after embodying a black avatar in virtual reality (Peck, Seinfeld, Aglioti, & Slater, 2013).

Perhaps fiction reading (and frequent immersion) could cultivate more empathy. Evidence shows that fiction readers have stronger empathy because they possess a stronger *theory of mind* (Mar, Oatley, & Peterson, 2009). These studies debunk the stereotype that fiction readers are socially awkward bookworms. In actuality, they have strong social skills because they can simulate other perspectives more easily. Nonfiction readers are the socially inept individuals (Mar et al., 2009).

But hey, at least we'll die with useless knowledge, right?

Narration

Wait, do you hear that? Do you hear a voice inside your head speaking these words? What is this voice? Is it *your* voice? *My* voice? Why does it exist? Let's demystify this inner speech.

First, this inner voice is your voice (with your accent). Researchers determined that conclusion through a clever limerick:

> There was a young runner from Bath,
> Who stumbled and fell on the path;
> She didn't get picked,
> As the coach was quite strict,
> So he gave the position to Garth.

Speakers from Southern England pronounce Garth like Bath, so they could read this limerick without any trouble. However, speakers from Northern England had trouble (Filik, & Barber, 2011). Therefore, inner speech is *your* voice.

Although you narrate with your own voice, you simulate other voices when characters are speaking. For example:

- Mary said that she was hungry.
- Mary said, "I'm hungry."

The second sentence is more vivid because it activates the auditory cortex (Yao, Belin, & Scheepers, 2012).

Not only do you simulate a voice, but you also simulate traits about that particular voice, like speaking rate.

- **Indirect.** Slowly, he looked around and said that he was grateful for their coming and that it was the end of the journey because it was unlikely that he would make it this time.
- **Direct.** Slowly, he looked around and said: "I'm grateful you're all here. This is the end of the journey because it is unlikely that I will make it this time."

People were slower to read the direct speech because they simulated a slower speaking voice (Yao & Scheepers, 2011).

In another study, people heard a conversation between two speakers; one spoke fast, while the other spoke slow. Then, they read a passage that was written by one of them. Reading speeds matched whichever speaker—fast or slow—supposedly wrote the passage (Alexander & Nygaard, 2008).

Dynamic Simulation in Film

Reading activates a simulation of the depicted events. Watching films will activate those simulations in real-time.

More importantly, I'll argue this claim: When films match your internal simulation, you immerse yourself into these films more vividly.

You might notice a problem, though. Don't you prefer simulations with seamless continuity? So then, why do films cut to new frames every few seconds? Why don't they display one seamless shot?

Here's the answer: Seamless continuity isn't *visual* continuity. You might be watching dialogue in a film, but then you notice a character looking in a particular direction. At this moment, you will wonder what this character is viewing, so you immerse yourself into this character to see their viewpoint. The director needs to cut to this character's viewpoint to match your unfolding simulation. This cut has a visual change, but it maintains seamless continuity because it depicts the next thing that you are expecting to see.

Plus, you don't view the world in a seamless fashion. You view the world by jumping between fixation points, a movement called an *eye saccade*. Even though films cut to a new shot every few seconds, you rarely notice them because these cuts are replicating your simulations (which include these cuts). From your perspective, everything is seamless.

Directors can also camouflage their cuts by matching your simulation, a technique that I call *editing continuity*. Here some examples.

| Shot 1 | Shot 2 | Shot 3 |

Figure 18D

Start with an Establishing Shot

Directors start with an "establishing shot" that depicts the room, characters, and overall context. Figure 18D depicts three shots in the beginning of a scene. The first shot displays the outside of an office building. Next, you see the overall layout of the room in which characters are speaking. Finally, you see a close-up shot of the screen that everybody is viewing.

If a director started with the close-up shot, you wouldn't be able to simulate the objects and characters outside of this frame. This confusion would make it harder to immerse yourself into the scene.

Overlap Cuts with Sound

Much like certain words (e.g., and, or, but) can help you bind new sentences into a simulation, sounds can accomplish this effect in film. You might hear characters speaking during the building shot in Figure 18D. At this point, you start simulating a conversation, and *bam*—the shot changes to a conversation. This new shot matches your unfolding simulation, so you remain blissfully unaware of the cut.

Summary

You read sentences by constructing mental images. Want to improve the clarity of your writing? Help readers integrate new information into their unfolding simulations. For example, use different types of seamless continuity:

- **Active:** Place subjects early in the sentence.
- **Descriptive:** Choose illustrative verbs.
- **Referential:** Start sentences from the last mental image.
- **Connective:** Bind sentences with connective words.
- **Expected:** Match their expected simulation.
- **Singular:** Constrain sentences to one possible image.
- **Temporal:** Group events within the same time frame.

You also immerse yourself into protagonists. Marketers advertise through stories because you *become* those characters—for example, smokers were less likely to smoke after feeling transported by an anti-smoking ad (Kaufman & Libby, 2012).

Other examples:

- **Speaking Rate From Reading.** Perhaps reading has dictated the speed in which you talk. Some writing systems are easier (and quicker) to read, and people speak faster in those languages (Gagl et al., 2018). You might speak at your current pace because of your reading speed.
- **Emotion and Intonation.** Suppose that somebody stole an object. If you wonder *who* stole the object, you'll be quicker to read a sentence that emphasizes the subject: JAMES stole the bracelet. If you wonder *what* was stolen, you'll be quicker to read a sentence that emphasizes the object: James stole

the BRACELET (Gross, Millett, Bartek, Bredell, & Winegard, 2014). In both cases, the capital letters match the intonation of your inner voice.

- **Chronological Sequence.** Information seems more serendipitous in a chronological sequence. In legal trials, witnesses usually testify in a random order, yet their testimony seems more credible when it follows the chronological sequence of events (Pennington & Hastie, 1986).

19
Time

THIS BOOK HAS thrown a lot of information at you. In this final chapter, we'll summarize the main principles.

Here's the key premise of the book: Sensory concepts fueled your knowledge of the world. You entered this world without knowing anything except the confusing mixture of sensory concepts. You took these primitive ideas—size, distance, motion—and you built your knowledge onto this foundation. Today, all abstract concepts are built with sensory frameworks.

In this chapter, we'll revisit every primitive to illustrate how these ideas are sculpting the abstract concept of time.

Size

Durations of time can be short or long because you conceptualize these ideas with a sensory concept of SIZE.

In fact, you believe that visually large stimuli appear on a screen for longer durations of time. Same with numbers: High numbers seem to appear for a longer duration than low numbers (Xuan, Zhang, He, & Chen, 2007; Oliveri et al., 2008).

Object

You group the sensory world into discrete objects, and you extend this behavior into the domain of time:

> . . . a person who visits a modern art gallery, a classic art gallery, the opera, and a symphony concert could either construe these as four distinct experiences or as two categories of experiences (art galleries and musical performances; Shah & Alter, 2014, p. 965).

You can also influence the perceived duration of time by placing boundaries in strategic places. You often see carpets in front of ATMs because these carpets expand the boundaries of the ATM, shortening the distance between people and their destination (see Figure 19A; Zhao et al., 2012). Waiting times seem shorter because it feels like you reached the ATM upon reaching the carpet.

Hospitals apply this technique, too: You wait in the general waiting room, and then you wait in the examination room. When you

Figure 19A

reach the examination room, you believe that you already reached your destination.

Have you ever wondered why return trips seem shorter? Return trips seem shorter because you conceptualize your home with a larger boundary, so you reach your home sooner:

> Since home is extremely familiar it enjoys a rich mental representation, and therefore, consumers may encode it as a relatively larger geographical area than the less familiar destination (Raghubir, Morwitz, & Chakravarti, 2011, p. 1).

Traveling to an unfamiliar place? You will arrive upon reaching the final destination itself. It's like waiting for an ATM without a carpet.

Some researchers oppose that explanation because this effect occurs while returning from unfamiliar routes (van de Ven, van Rijswijk, & Roy, 2011). So, perhaps you could blame an additional sensory experience. Consider a rubber band. Pulling a rubber band requires effort, but what happens when you let go? Both sides snap together very quickly. Perhaps you conceptualize travel with a similar metaphor. When you travel away from home, you move away from your default state (like pulling a rubber band). This motion feels slow and effortful. However, when you return home, you move toward your default state. Much like a rubber band, you perceive this returning motion to be quicker and easier.

Shape

I exercise every weekday, but I noticed that my motivation disappears toward the end of my workout. If three sets are remaining, I'm tempted to forgo the last one or two sets. I'm not physically tired. I have the capability to do them. But the sets—themselves—feel less important, as if I've already done enough that day. Do you ever feel that way? I have an idea why.

When I conceptualize my workout, I conceptualize an hour that I need to "fill" with exercise. Aha . . . I conceptualize this duration as a CONTAINER. Whenever I fill a physical container, such as a glass of water, I don't fill the maximum capacity; I leave room at the top so that it doesn't spill over. If I conceptualize TIME as a CONTAINER, I might enact the same behavior. I'll neglect the final section of my workout because the container seems filled enough.

Location

The past is on the left, and the future is on the right. But this perception is malleable—within 5 minutes of reading reversed text, people conceptualized time in the reverse direction (Casasanto & Bottini, 2014a).

Sound

Words about the past seem louder in your left ear, while words about the future seem louder in your right ear (Lakens, Semin, & Garrido, 2011).

Distance

You can see the details of objects that are close to you, but you can only see the gist of distant objects. You insert this experience into time: You focus on the details of nearby events, but you focus on the gist of distant events (D'Argembeau & Van der Linden, 2004).

Enrolling in courses? You prefer courses with desirability (e.g., quality of professor) if the deadline is far away, but you prefer courses with feasibility (e.g., convenient location) if the deadline is soon (Fujita, Eyal, Chaiken, Trope, & Liberman, 2008).

Color

Your mental images are detailed and colorful for events that are happening soon, but you generate black-and-white simulations for events in the distant future (see Chapter 6 for details).

Motion

You conceptualize time with sensory motion:

> There is an area in the visual system of our brains dedicated to the detection of motion. There is no such area for the detection of global time. That means that motion is directly perceived and is available for use as a source domain by our metaphor systems (Lakoff & Johnson, 1999, p. 140).

You conceive this motion in two ways: You move toward the future, or the future moves toward you.

Your choice—Moving Observer vs. Moving Time—depends on the perception that is most adaptive. Suppose that you recall an unpleasant memory in the past. You will conceptualize Moving Observer so that you move away from this bad memory. However, if you recall a pleasant memory, you will conceptualize Moving Time so that you cling to the past.

The opposite happens with future events: With pleasant events in the future, you adopt Moving Observer to move toward them; with unpleasant events in the future; you adopt Moving Time to stall your movement (Lee & Ji, 2014).

Orientation

Western cultures place the future in front of you: People shift forward when they imagine future events, but they shift backward for past events (Miles, Nind, & Macrae, 2010).

Even within cultures, people can view time differently based on personality or political views. Conservatives are focused on the past, while liberals are focused on the future. Conservative websites (e.g., Drudge Report) use the past tense more often, while liberal websites (e.g., Huffington Post) use the future tense more often (Robinson, Cassidy, Boyd, & Fetterman, 2015). You find similar effects in presidential speeches: Republicans reference the past more often, while Democrats reference the future more often.

Emotion

Future stimuli feel more intense. Consider these events:

- ▶ Positive (e.g., Thanksgiving Day)
- ▶ Negative (e.g., annoying noise)
- ▶ Routine (e.g., menstruation)
- ▶ Hypothetical (e.g., all-expenses-paid ski vacation)

All of those events felt more intense when they would be occurring in the future, compared to the equidistant past (Van Boven & Ashworth, 2007).

Future events feel more emotional, which can distort your perception of morality. People perceived more wrongdoing when they believed that Coca-Cola was thinking about algorithmically raising prices on hot days. If those policies were already in place, they didn't care as much (Caruso, 2010).

In another study, researchers asked Harvard students to imagine spending 5 hours on data entry—either 1 month in the past or 1 month in the future. When asked their desired compensation, students in the future condition asked for 101% more money ($125.04 vs. $62.20; Caruso, Gilbert, & Wilson, 2008). Students placed more value on their future work.

Sure, a $60 difference seems minor, but the same effect happens on a larger scale. Compare these two versions of a legal trial in which a woman was injured by a drunk driver:

- **Version A.** The woman is fully recovered, but she underwent 6 months of painful rehabilitation.
- **Version B.** The women will fully recover, but she will need to undergo 6 months of painful rehabilitation.

Participants awarded $2.5 million in Version A, yet $3.6 million in Version B (Caruso et al., 2008). They awarded an additional $1 million in compensation when the rehabilitation was occurring in the future because it seemed more arduous.

Physiology

Time moves faster when you feel distracted. Ever wonder why you see mirrors near elevators? Mirrors distract you from noticing each second that ticks away (Gorn, Chattopadhyay, Sengupta, & Tripathi, 2004).

Likewise, time moves slower during high arousal (e.g., robbery, speech, interview) because you are more aware and cognizant. You notice each second that ticks away (Droit-Volet, Mermillod, Cocenas-Silva, & Gil, 2010).

Warm colors, which increase arousal, can slow your perception of time:

> ... the uniforms of the checkout employees might influence perceived ease and time spent during the transaction ... a store like Target, with its almost overwhelming, saturated red atmosphere at the checkout area, may need to reconsider its interior color choices (Labrecque, 2010, p. 30).

Cool colors can speed this perception: You're more likely to wait for something to download if the background color is blue (Gorn et al., 2004).

People

You predict your future mindset by inserting your present self into your future self. Hungry people bid more money on future food because they were biased by their current level of hunger (Fisher & Rangel, 2014). Men who felt sexually aroused indicated a greater likelihood of aggressive sexual behavior in the future (Loewenstein, Nagin, & Paternoster, 1997).

Putting It All Together

Let's walk through a real-world example to see how these primitive concepts are influencing your perception and behavior.

Suppose that you live in Florida, and you're taking a trip to New York. Each primitive will influence your perception of these events.

Buying Tickets

You are buying tickets for the flight. You typically stick with economy, but for some reason, you snag the extra comfort this time.

But wait . . . did you really need the extra space? Or can you blame your obnoxiously tight cubicle at work?

If you bought these tickets while feeling confined in the present moment, you inserted this need into your future self. You preferred upgraded seating because you were craving the extra space in your cubicle. Luckily, ticket kiosks in airports are usually located in open areas; otherwise, you might see an uptick in first class upgrades.

Perhaps you could also blame spatial confinement. If you feel spatially confined in a cubicle, you'd want to reclaim your freedom by making decisions that are more atypical (see Chapter 8).

Figure 19B. How I buy flight tickets.

Waiting for the Trip

You bought tickets, but the trip is still 3 months away. When something is far away, you can't see the details (so you focus on the gist). At this moment, you will be focusing on the desirable and high-level aspects of the trip (e.g., what you'll do). As you move closer, you will start focusing on the feasible and concrete details of the trip (e.g., how you'll do it).

During the Flight

You are flying from Florida to New York. This northern route will seem longer and more effortful because it feels like you are moving upward, a direction that is hindered by gravity.

This trip also seems slower because of higher arousal:

> While trips home may be associated with pleasant thoughts (e.g., looking forward to watching TV, relaxing, and having dinner), trips away from home could sometimes be associated with unpleasant thoughts, worry, or anxiety (e.g., worrying while traveling to a doctor, or to school because of an impending exam; Raghubir et al., 2011, p. 5).

Baggage Collection

Arriving passengers collect baggage by walking to a new area. It seems annoying, but this walk is less annoying than waiting at the gate.

Suppose that airports need 10 minutes to deliver your bags. If you wait at the gate, you'll be fixated on each second that ticks away. Plus, you won't be moving forward spatially, so it feels like you aren't moving forward in time. If you walk 8 minutes, and then wait

2 minutes for your bags, you will be moving forward (and distracted by the walk).

Summary

Time is an abstract concept. In order to understand and conceptualize this idea, you inserted a sensory framework. Whenever you conceptualize time, you are activating this primitive foundation.

Usually, authors begin a chapter by defining a topic. We described the sensory components of time, but—ultimately—what *is* time? Can we define this idea without sensory concepts? What if the world didn't contain space? Or distance? Or motion? Would time still exist?

And how many other abstract concepts exist? Are we surrounded by a sea of important abstract ideas, yet we remain blind to these ideas because we don't possess the necessary primitives to understand them?

I honestly don't know. But it's time to end this book.

Conclusion

PHEW, it was a long journey, my friend—but we finally made it.

Many scientists perceive the field of *embodied cognition*—the topic of this book—to be a joke. Some findings are unusual and surprising, so many academics have distanced themselves from this field because of the controversy and skepticism.

After immersing myself into this research, I can confidently say that I'm a firm believer: Sensory concepts sculpted our knowledge and thinking. Hopefully this book restores the credibility of this field. And hopefully my endeavor hasn't labeled me another conspiracy junkie.

In this final section, I'll expand on applications that we never discussed, and I'll describe the future outlook of these concepts.

Education

Children need spatial knowledge to learn abstract concepts, like math. Some parents mock "fluffy" skills like art or music, asking: *When will my child ever use this?* Well, now you know. Fluffy skills can expand the sensory basis of knowledge, helping children learn not-so-fluffy skills later.

Policy makers can use these ideas to improve school curriculums.

Schools should be encouraging a liberal arts education with a diversity of topics, rather than confining students to a narrow focus.

Music

I listen to metal . . . weird metal. Two seconds ago, my Spotify playlist just shuffled to the song Trollhammeren by Finntroll. Listen to that song if you want to experience my eclectic tastes (earplugs not provided).

My musical tastes might seem unusual for a research nerd . . . but are they? In this section, I'll argue for *canonical music*: You prefer music that matches your expectations.

Where do you derive these expectations? Perhaps three sources: listener convergence, musical patterns, and body movement.

Listener Convergence

You gravitate toward pets that resemble your own face and body structure. I suspect that you also prefer music that matches your identity. Evidence shows that you can accurately judge somebody based on their music preferences alone (Adrian, 2010).

My outward appearance doesn't fit the stereotype of a metal listener, but you need to dig deeper than *genre* of music:

> . . . relying solely on genre labels makes it hard to know which aspects of music influence preferences. Listeners could be drawn to auditory and psychological properties that are intrinsic to the music, such as timbre, pitch, or intensity. These in turn, especially when combined with lyrics, can give rise to specific emotional reactions to the music that are genre-independent (Rentfrow et al., 2012, p. 4).

Consider the concept of SOFT. Objects can feel physically soft, yet certain music can also be characterized as soft. Interestingly, you prefer congruence in these domains: People evaluated products more favorably while standing on a soft floor listening to soft music (Imschloss & Kuehnl, 2017).

More importantly, these concepts also occur in people. You might prefer music that matches the primitive traits of your identity:

- **Volume.** In the early 90's, evidence showed that women preferred music with lower volumes (Kellaris & Rice, 1993). I'd be curious to see a replication today. Over the past few decades, cultures have been pushing more attention on the importance of gender equality. Perhaps women preferred lower volumes of music because this trait resembled the metaphorical suppression of their voice. If women are starting to feel like their voice is being heard, perhaps they prefer higher volumes today.
- **Bass.** Males prefer stronger bass (Colley, 2008). Perhaps this bass resembles the low-pitched tonality of their voice.
- **Pitch.** Higher pitches (and ascending tones) seem happier, faster, and brighter (Collier & Hubbard, 1998). Perhaps happier people gravitate toward that music.

I opened this section by mentioning my weird taste in metal, but maybe it's not so weird after all. Certain concepts appear in that music and my personality.

- **Speed.** I like fast music, and I also like speed and progress in all facets of my life (slow walkers are the worst). Even in my writing, I try to eliminate all fluff so that we move quickly from concept to concept.
- **Distinctive.** I like distinctive music, and I also like to be distinctive in everything I do. You might have noticed that this book is different than most books.

▶ **Lighthearted.** If you listen to the song that I mentioned earlier, you'll hear a heavy and intense song with a sense of lightheartedness. That quality is a metaphor for this book (and my other work). I convey intense research in a lighthearted way.

My arguments are anecdotal, but hopefully researchers can verify these ideas in empirical studies. In the meantime, let's see another factor behind our music preferences.

Musical Patterns

You prefer objects with "typical" traits. And music follows a "typical" pattern. Even babies can predict upcoming beats in music (Winkler, Háden, Ladinig, Sziller, & Honing, 2009).

Eventually, you become so proficient that you can predict upcoming sounds, and you feel good upon hearing the congruence. Perhaps this notion can explain why repetitive songs are more likely to reach the #1 spot in the Top 40 (Nunes, Ordanini, & Valsesia, 2015).

You can also search for a comedic song called "The 4 Chords Song" that exposes the same rhythmic pattern in many popular hits. If you frequently hear music with the same rhythm, you develop a canonical rhythm (and you prefer music that matches this pattern). Ironically, the 4 Chords Song—which incorporated that pattern—became a hit itself. Ironic indeed.

Motor Simulation

Which came first: music or dancing? Seems obvious . . . music, right? Well, not so fast.

Any sound—banging, clapping, stomping—requires movement. It's impossible to produce any sound without a sense of movement. Even humming requires movement in the vocal cords.

Movements are part of those sounds. Thus, whenever you hear

a sound, you simulate the necessary movement. Researchers tested this concept in people with apraxia, a condition that impairs body movement. If people couldn't move their facial muscles, they had trouble distinguishing a cough from a yawn; if people couldn't move their limbs, they had trouble distinguishing a finger snap from a hand clap (Pazzaglia, Pizzamiglio, Pes, & Aglioti, 2008). You identify sounds by simulating the movements that produce these sounds.

So, why do you dance? While listening to music, you feel an urge to move because you are simulating the movements that are producing these sounds. You might even tailor your body movements to match the metaphors of the sound, such as moving your body upward during high pitches.

You might also prefer music that matches a canonical movement. Boxers might prefer music with intense and punctuated beats because they perform strong and punctuated movements with their arms. Practitioners of tai chi might prefer music with soft and fluid rhythm.

Art & Design

Right now, find an object nearby. Whatever you find—laptop, table, wallet—somebody probably "designed" it.

This book unraveled your perception of the world, yet the world is flooded with design. If designers understand the implications of sensory concepts, they can improve their designs (and the world around us).

When I started writing this book, I planned to write a book on design; the original title was *Methods of Design*. Even though I branched into other topics, this book is still a great textbook for designers. Not only does it unify separate fields of design into a common set of primitives, but it also validates the scientific value of design. Elements of design (e.g., size, location, shape) dictate your

Figure 20A. Adapted from Krishna (2013)

perception of the world. Scientists should be studying these concepts with the academic rigor they deserve.

Motor Simulation

You just learned that you simulate the movements that produce sound. The same effect occurs in artwork. When you see an abstract painting, you simulate the hand movements that produced the painting (Umilta et al., 2012).

You see a similar effect in Figure 20A. Don't the cookies seem lighter at the top of a package, as if they're diet cookies? Sure enough, dieters prefer those designs (Deng & Kahn, 2009). When you see food at the top of a package, you simulate a lifting motion that would have placed the food in that location (and thus it seems lighter).

Golden Ratio

Our framework can solve the mystery of the *golden ratio*. Stimuli seem more beautiful when they adhere to a ratio of 1.618 (see Figure 20B). You also prefer spirals that expand by that ratio.

A   **B**

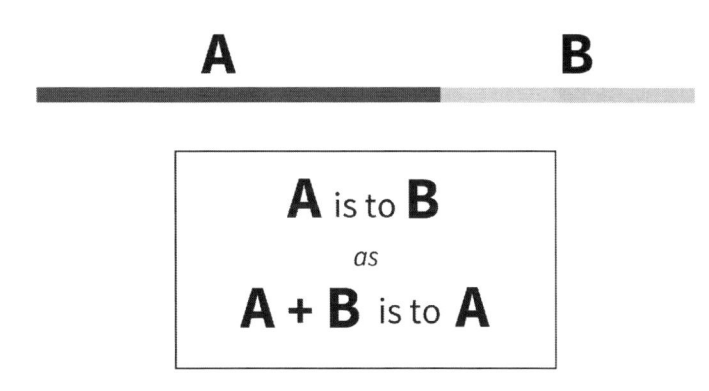

A is to **B**
as
A + B is to **A**

Figure 20B

So, why is that ratio beautiful?

Earlier you learned that nature is evolving toward adaptive patterns, like symmetry, because these patterns transport energy more effectively. Golden ratios are another adaptive pattern. In leaves, this ratio creates a spiral in which all leaves capture sunlight without blocking other leaves (Omotehinwa & Ramon, 2013). In biology, this ratio provides the optimal pump structure for a human heart (Henein et al., 2011).

Golden ratios can be found in many aspects of nature because they were adaptive in evolution, and—as a result—you developed a canonical ratio for many stimuli. You prefer this proportion in many contexts because it matches the typical ratio.

Replication Crisis

Solid research is replicable. If somebody publishes an experiment to support a claim, an identical experiment should produce the same result. If it doesn't replicate, the claim isn't true.

In recent years, social scientists have been running studies of past experiments; and the results are somewhat frightening. Among a large sample of replications, only 36% were successful (Open Science Collaboration, 2015).

Amidst this crisis, researchers have focused their attacks on "social priming" because of a popular study. When exposed to words about the elderly—bingo, retired, Florida—people walked out of the room slower (Bargh, Chen, & Burrows, 1996). This study garnered 5,000 citations, which is an impressive feat in academia. Unfortunately, backlash started when researchers failed to replicate these effects (Doyen, Klein, Pichon, & Cleeremans, 2012). Those failures prompted Daniel Kahneman, a respected scientist and Nobel prize winner, to write an open letter criticizing the state of priming. His letter solidified this field (along with embodied cognition) as a poster child for the replication crisis.

Why can't we replicate past findings? Sometimes it's outright fraud. According to a study, roughly 2% of scientists have admitted to falsifying data (Fanelli, 2009). And who knows if that study was part of the 2%.

Even more problematic, many scientists evaluate findings with a 95% significance level. If you measure 100 variables that influence eating behavior—food color, seating arrangement, and more—roughly 5 variables will show a "significant" impact based on random chance alone. You could publish a paper explaining why these variables influence eating behavior, while neglecting to mention the other variables that you measured.

Good researchers don't follow those practices, but how can you distinguish good data from bad data? Reputable journals will retract studies after discovering those manipulative practices, but—by that point—those studies are already ingrained into the theoretical literature. While writing this book, I noticed that journals were retracting papers from a popular researcher in food psychology because of bad practices. I removed those citations, but I don't know whether other studies are suffering from the same flaws. Scientists need to solve this problem.

Some researchers argue that the "crisis" is less severe than it sounds because we have *conceptual replications*. Consider the study with elderly priming. Even though future studies failed to replicate the effect of slower walking, priming has been supported through an abundance of different methodologies. We can safely ignore a failed replication of priming because we have decades of studies that support priming in different contexts.

Likewise, this book provided an onslaught of studies that illustrate how sensory concepts shape our knowledge. Embodied cognition makes some radical claims, and I believe that radical claims require a radical amount of evidence. So, I crammed this book with research to showcase the overwhelming aggregate of evidence. Some studies could, indeed, be flawed—but it's hard to dismiss thousands of studies pointing to the same conclusion.

How Impactful Are These Concepts?

This book offered a lot of applications, but how much do they really impact you? It depends. Some concepts are stronger than others:

- ▶ **Small Impact.** If you see your wallet near some rubber bands, would you really buy something because of your flexible spending? You might . . . but the impact would be minimal.
- ▶ **Big Impact.** If you see a stimulus that matches your expectations, you evaluate it more favorably. This effect influences your decisions every day.

You also need to consider the macro level. Suppose that the weight of your credit card increases spending by 0.01%. Whoop-dee-doo, right? Well, across a population, that tiny bias could influence hundreds of millions of dollars.

And that's one principle. By understanding all of the sensory concepts, we can fix small biases that are generating a large impact on the population.

Final Thoughts

You are surrounded by sensory concepts that shaped everything about you. Your knowledge. Your perception. Your behavior. *Everything*. I hope that you see the world differently moving forward.

If this book was your first experience with me, then I hope you enjoyed our journey. And I hope to see you again. Your brain is a fascinating entanglement that we will never fully understand.

We will always have more to unravel.

Next Step

I RARELY POST on social media, but I occasionally send an email newsletter when I think of something interesting to share. I try to make every newsletter worthwhile to read. You can subscribe on my website.

www.NickKolenda.com

Other Books

YOU CAN CHECK my Amazon page to see my current list of books. At the moment, I've written two other books:

- **Methods of Persuasion (2012).** I wrote this book in my early twenties, near the beginning of my venture into research, but I think it's a fun summary of existing research on persuasion.
- **Imagine Reading This Book (2020).** *The Tangled Mind* can be a little intimidating for a broad audience, so I wrote a follow-up book—*Imagine Reading This Book*—to illustrate some of the core ideas in a more digestible format. But it's still worthwhile to read even if you've read *The Tangled Mind*. You'll discover the key mechanism that determines every decision, and you'll learn how to apply this principle to motive yourself or others.

References

Aaker, J., Vohs, K. D., & Mogilner, C. (2010). Nonprofits are seen as warm and for-profits as competent: Firm stereotypes matter. *Journal of Consumer Research*, 37(2), 224–237.

Ackerman, J. M., Nocera, C. C., & Bargh, J. A. (2010). Incidental haptic sensations influence social judgments and decisions. *Science*, 328(5986), 1712–1715.

Adrian, C. (2010). Individual differences in musical taste. *American Journal of Psychology*, 123(2), 199–208.

Alards-Tomalin, D., Walker, A. C., Nepon, H., & Leboe-McGowan, L. C. (2017). Dual-task interference effects on cross-modal numerical order and sound intensity judgments: the more the louder? *The Quarterly Journal of Experimental Psychology*, 70(9), 1943–1963.

Alexander, G. M. (2003). An evolutionary perspective of sex-typed toy preferences: Pink, blue, and the brain. *Archives of Sexual Behavior*, 32(1), 7–14.

Alexander, J. D., & Nygaard, L. C. (2008). Reading voices and hearing text: Talker-specific auditory imagery in reading. *Journal of Experimental Psychology: Human Perception and Performance*, 34(2), 446.

Alter, A. L., & Oppenheimer, D. M. (2006). Predicting short-term stock fluctuations by using processing fluency. *Proceedings of the National Academy of Sciences*, 103(24), 9369–9372.

Alter, A. L., & Oppenheimer, D. M. (2008). Effects of fluency on psychological distance and mental construal (or why New York is a large city, but New York is a civilized jungle). *Psychological Science*, 19(2), 161–167.

Alter, A. L., & Oppenheimer, D. M. (2009). Uniting the tribes of fluency to form a metacognitive nation. *Personality and Social Psychology Review*, 13(3), 219–235.

Appel, M. (2011). A story about a stupid person can make you act stupid (or smart): Behavioral assimilation (and contrast) as narrative impact. *Media Psychology*, 14(2), 144–167.

Ariga, A., & Watanabe, K. (2009). What is special about the index finger? The index finger advantage in manipulating reflexive attentional shift. *Japanese Psychological Research*, 51(4), 258–265.

Arnheim, R. (1974). Art and visual perception: A psychology of the creative eye. Berkeley and Los Angeles, CA: University of California Press.

Arrighi, R., Cartocci, G., & Burr, D. (2011). Reduced perceptual sensitivity for biological motion in paraplegia patients. *Current Biology*, 21(22), R910–R911.

ASIRT (2018). Annual Global Road Crash Statistics. Retrieved from https://www.asirt.org/safe-travel/road-safety-facts/

Asutay, E., & Västfjäll, D. (2012). Perception of loudness is influenced by emotion. *PloS One*, 7(6), e38660.

Atalay, A. S., Bodur, H. O., & Rasolofoarison, D. (2012). Shining in the center: Central gaze cascade effect on product choice. *Journal of Consumer Research*, 39(4), 848–866.
Attrill, M. J., Gresty, K. A., Hill, R. A., & Barton, R. A. (2008). Red shirt colour is associated with long-term team success in English football. *Journal of Sports Sciences*, 26(6), 577–582.
Bagchi, R., & Cheema, A. (2012). The effect of red background color on willingness-to-pay: The moderating role of selling mechanism. *Journal of Consumer Research*, 39(5), 947–960.
Bailey, J. M., Gaulin, S., Agyei, Y., & Gladue, B. A. (1994). Effects of gender and sexual orientation on evolutionarily relevant aspects of human mating psychology. *Journal of Personality and Social Psychology*, 66(6), 1081.
Bakhtiari, G. (2015). The role of fluency in oral approach and avoidance. Doctoral dissertation.
Bakhtiari, G., Körner, A., & Topolinski, S. (2016). The role of fluency in preferences for inward over outward words. *Acta Psychologica*, 171, 110–117.
Bal, P. M., & Veltkamp, M. (2013). How does fiction reading influence empathy? An experimental investigation on the role of emotional transportation. *PloS One*, 8(1), e55341.
Balas, B., & Saville, A. (2015). N170 face specificity and face memory depend on hometown size. *Neuropsychologia*, 69, 211–217.
Banerjee, P., Chatterjee, P., & Sinha, J. (2012). Is it light or dark? Recalling moral behavior changes perception of brightness. *Psychological Science*, 23(4), 407–409.
Bar-Anan, Y., Liberman, N., & Trope, Y. (2006). The association between psychological distance and construal level: evidence from an implicit association test. *Journal of Experimental Psychology: General*, 135(4), 609.
Bar-Anan, Y., Liberman, N., Trope, Y., & Algom, D. (2007). Automatic processing of psychological distance: evidence from a Stroop task. *Journal of Experimental Psychology: General*, 136(4), 610.
Bar-Haim, Y., Ziv, T., Lamy, D., & Hodes, R. M. (2006). Nature and nurture in own-race face processing. *Psychological Science*, 17(2), 159–163.
Bar, M., & Neta, M. (2006). Humans prefer curved visual objects. *Psychological Science*, 17(8), 645–648.
Bar, M., Neta, M., & Linz, H. (2006). Very first impressions. *Emotion*, 6(2), 269.
Barber, N. (2011). A cross-national test of the uncertainty hypothesis of religious belief. *Cross-Cultural Research*, 45(3), 318–333.
Bargh, J. A., & Shalev, I. (2012). The substitutability of physical and social warmth in daily life. *Emotion*, 12(1), 154.
Bargh, J. A., Chen, M., & Burrows, L. (1996). Automaticity of social behavior: Direct effects of trait construct and stereotype activation on action. *Journal of Personality and Social Psychology*, 71(2), 230.
Baron-Cohen, S., & Wheelwright, S. (2004). The empathy quotient: an investigation of adults with Asperger syndrome or high functioning autism, and normal sex differences. *Journal of Autism and Developmental Disorders*, 34(2), 163–175.
Barrett, L. F. (2017). How emotions are made: The secret life of the brain. Houghton Mifflin Harcourt.
Barsalou, L. (2003). Situated simulation in the human conceptual system. *Language and Cognitive Processes*, 18(5–6), 513–562.
Barsalou, L. W. (1999). Perceptual symbol systems. *Behavioral and Brain Sciences*, 22(4), 577–660.
Bastian, B., Jetten, J., & Fasoli, F. (2011). Cleansing the soul by hurting the flesh: The guilt-reducing effect of pain. *Psychological Science*, 22(3), 334.

Bateson, M., Nettle, D., & Roberts, G. (2006). Cues of being watched enhance cooperation in a real-world setting. *Biology Letters*, 2(3), 412–414.

Bejan, A., & Lorente, S. (2010). The constructal law of design and evolution in nature. *Philosophical Transactions of the Royal Society B: Biological Sciences*, 365(1545), 1335–1347.

Bem, S. L. (1981). Gender schema theory: A cognitive account of sex typing. *Psychological Review*, 88(4), 354.

Bereczkei, T., Gyuris, P., Koves, P., & Bernath, L. (2002). Homogamy, genetic similarity, and imprinting; parental influence on mate choice preferences. *Personality and Individual Differences*, 33(5), 677–690.

Bergen, B. K. (2012). *Louder than words: The new science of how the mind makes meaning*. New York, NY: Basic Books.

Bertrand, M., & Mullainathan, S. (2004). Are Emily and Greg more employable than Lakisha and Jamal? A field experiment on labor market discrimination. *American Economic Review*, 94(4), 991–1013.

Bhalla, M., & Proffitt, D. R. (1999). Visual–motor recalibration in geographical slant perception. *Journal of Experimental Psychology: Human perception and performance*, 25(4), 1076.

Birch, S. A., & Bloom, P. (2004). Understanding children's and adults' limitations in mental state reasoning. *Trends in Cognitive Sciences*, 8(6), 255–260.

Blickenstaff, C. J. (2005). Women and science careers: leaky pipeline or gender filter? *Gender and Education*, 17(4), 369–386.

Blogowska, J., Lambert, C., & Saroglou, V. (2013). Religious prosociality and aggression: It's real. *Journal for the Scientific Study of Religion*, 52(3), 524–536.

Bóo, F. L., Rossi, M. A., & Urzúa, S. S. (2013). The labor market return to an attractive face: Evidence from a field experiment. *Economics Letters*, 118(1), 170–172.

Boroditsky, L., & Ramscar, M. (2002). The roles of body and mind in abstract thought. *Psychological Science*, 13(2), 185–189.

Boroditsky, L., Schmidt, L. A., & Phillips, W. (2003). Sex, syntax, and semantics. *Language in mind: Advances in the study of language and thought*, 61–79.

Bremner, A. J., Caparos, S., Davidoff, J., de Fockert, J., Linnell, K. J., & Spence, C. (2013). "Bouba" and "Kiki" in Namibia? A remote culture make similar shape–sound matches, but different shape–taste matches to Westerners. *Cognition*, 126(2), 165–172.

Bronstad, P. M., & Russell, R. (2007). Beauty is in the 'we'of the beholder: Greater agreement on facial attractiveness among close relations. *Perception*, 36(11), 1674–1681.

Bruner, J. S., & Goodman, C. C. (1947). Value and need as organizing factors in perception. *Journal of Abnormal and Social Psychology*, 42(1), 33.

Brunyé, T. T., Ditman, T., Mahoney, C. R., Augustyn, J. S., & Taylor, H. A. (2009). When you and I share perspectives: Pronouns modulate perspective taking during narrative comprehension. *Psychological Science*, 20(1), 27–32.

Brunyé, T. T., Gardony, A., Mahoney, C. R., & Taylor, H. A. (2012). Body-specific representations of spatial location. *Cognition*, 123(2), 229–239.

Brunyé, T. T., Mahoney, C. R., Gardony, A. L., & Taylor, H. A. (2010). North is up (hill): Route planning heuristics in real-world environments. *Memory & Cognition*, 38(6), 700–712.

Burger, J. M., Messian, N., Patel, S., Del Prado, A., & Anderson, C. (2004). What a coincidence! The effects of incidental similarity on compliance. *Personality and Social Psychology Bulletin*, 30(1), 35–43.

Burger, J. M., Sanchez, J., Imberi, J. E., & Grande, L. R. (2009). The norm of reciprocity as an internalized social norm: Returning favors even when no one finds out. *Social Influence*, 4(1), 11–17.

Burson, K. A., Larrick, R. P., & Lynch Jr, J. G. (2009). Six of one, half dozen of the other: Expanding and contracting numerical dimensions produces preference reversals. *Psychological Science*, 20(9), 1074–1078.

Bushong, B., King, L. M., Camerer, C. F., & Rangel, A. (2010). Pavlovian processes in consumer choice: The physical presence of a good increases willingness-to-pay. *American Economic Review*, 100(4), 1556–71.

Buss, D. M., Larsen, R. J., Westen, D., & Semmelroth, J. (1992). Sex differences in jealousy: Evolution, physiology, and psychology. *Psychological Science*, 3(4), 251–256.

Butterfield, M. E., Hill, S. E., & Lord, C. G. (2012). Mangy mutt or furry friend? Anthropomorphism promotes animal welfare. *Journal of Experimental Social Psychology*, 48(4), 957–960.

Cacioppo, J. T., & Patrick, W. (2008). Loneliness: Human nature and the need for social connection. WW Norton & Company.

Caruso, E. M. (2010). When the future feels worse than the past: A temporal inconsistency in moral judgment. *Journal of Experimental Psychology: General*, 139(4), 610.

Caruso, E. M., Gilbert, D. T., & Wilson, T. D. (2008). A wrinkle in time: Asymmetric valuation of past and future events. *Psychological Science*, 19(8), 796–801.

Caruso, E. M., Mead, N. L., & Balcetis, E. (2009). Political partisanship influences perception of biracial candidates' skin tone. *Proceedings of the National Academy of Sciences*, 106(48), 20168–20173.

Caruso, E. M., Van Boven, L., Chin, M., & Ward, A. (2013). The temporal Doppler effect: When the future feels closer than the past. *Psychological Science*, 24(4), 530–536.

Casasanto, D. (2008). Similarity and proximity: When does close in space mean close in mind? *Memory & Cognition*, 36(6), 1047–1056.

Casasanto, D. (2009). Embodiment of abstract concepts: good and bad in right-and left-handers. *Journal of Experimental Psychology: General*, 138(3), 351.

Casasanto, D., & Bottini, R. (2014a). Mirror reading can reverse the flow of time. *Journal of Experimental Psychology: General*, 143(2), 473.

Casasanto, D., & Bottini, R. (2014b). Spatial language and abstract concepts. *Wiley Interdisciplinary Reviews: Cognitive Science*, 5(2), 139–149.

Casasanto, D., & Dijkstra, K. (2010). Motor action and emotional memory. *Cognition*, 115(1), 179–185.

Casasanto, D., & Jasmin, K. (2010). Good and bad in the hands of politicians: Spontaneous gestures during positive and negative speech. *PloS One*, 5(7), e11805.

Casasola, M., Wilbourn, M. P., & Yang, S. (2006). Can English-learning toddlers acquire and generalize a novel spatial word? *First Language*, 26(2), 187–205.

Chae, B., & Hoegg, J. (2013). The future looks "right": Effects of the horizontal location of advertising images on product attitude. *Journal of Consumer Research*, 40(2), 223–238.

Chapman, E., Baron-Cohen, S., Auyeung, B., Knickmeyer, R., Taylor, K., & Hackett, G. (2006). Fetal testosterone and empathy: evidence from the empathy quotient (EQ) and the "reading the mind in the eyes" test. *Social Neuroscience*, 1(2), 135–148.

Chapman, H. A., & Anderson, A. K. (2014). Trait physical disgust is related to moral judgments outside of the purity domain. *Emotion*, 14(2), 341.

Chapman, H. A., Kim, D. A., Susskind, J. M., & Anderson, A. K. (2009). In bad taste: Evidence for the oral origins of moral disgust. *Science*, 323(5918), 1222–1226.

Chartrand, T. L., & Bargh, J. A. (1999). The chameleon effect: the perception–behavior link and social interaction. *Journal of Personality and Social Psychology*, 76(6), 893.

Chatterjee, A., Southwood, M. H., & Basilico, D. (1999). Verbs, events and spatial representations. *Neuropsychologia*, 37(4), 395–402.

Chen, R. P., Wan, E. W., & Levy, E. (2017). The effect of social exclusion on consumer preference for anthropomorphized brands. *Journal of Consumer Psychology*, 27(1), 23–34.

Cialdini, R. B. (2007). Influence: The psychology of persuasion. New York, NY: Collins.

Cialdini, R. B., Brown, S. L., Lewis, B. P., Luce, C., & Neuberg, S. L. (1997). Reinterpreting the empathy–altruism relationship: When one into one equals oneness. *Journal of Personality and Social Psychology*, 73(3), 481.

Cialdini, R. B., Schaller, M., Houlihan, D., Arps, K., Fultz, J., & Beaman, A. L. (1987). Empathy-based helping: Is it selflessly or selfishly motivated? *Journal of Personality and Social Psychology*, 52(4), 749.

Cian, L., Krishna, A., & Elder, R. S. (2015). A sign of things to come: Behavioral change through dynamic iconography. *Journal of Consumer Research*, 41(6), 1426–1446.

Cian, L., Krishna, A., & Schwarz, N. (2015). Positioning rationality and emotion: Rationality is up and emotion is down. *Journal of Consumer Research*, 42(4), 632–651.

Claus, B., & Kelter, S. (2006). Comprehending narratives containing flashbacks: Evidence for temporally organized representations. *Journal of Experimental Psychology: Learning, Memory, and Cognition*, 32(5), 1031.

Colley, A. (2008). Young People's Musical Taste: Relationship With Gender and Gender-Related Traits 1. *Journal of Applied Social Psychology*, 38(8), 2039–2055.

Collier, W. G., & Hubbard, T. L. (1998). Judgments of happiness, brightness, speed and tempo change of auditory stimuli varying in pitch and tempo. *Psychomusicology: A Journal of Research in Music Cognition*, 17(1–2), 36.

Coulter, K. S. (2007). The effects of digit-direction on eye movement bias and price-rounding behavior. *Journal of Product & Brand Management*, 16(7), 501–508.

Coulter, K. S., & Coulter, R. A. (2005). Size does matter: The effects of magnitude representation congruency on price perceptions and purchase likelihood. *Journal of Consumer Psychology*, 15(1), 64–76.

Coulter, K. S., & Coulter, R. A. (2010). Small sounds, big deals: phonetic symbolism effects in pricing. *Journal of Consumer Research*, 37(2), 315–328.

Coulter, K. S., & Norberg, P. A. (2009). The effects of physical distance between regular and sale prices on numerical difference perceptions. *Journal of Consumer Psychology*, 19(2), 144–157.

Coulter, K. S., Choi, P., & Monroe, K. B. (2012). Comma N'cents in pricing: The effects of auditory representation encoding on price magnitude perceptions. *Journal of Consumer Psychology*, 22(3), 395–407.

Crawford, L., Margolies, S. M., Drake, J. T., & Murphy, M. E. (2006). Affect biases memory of location: Evidence for the spatial representation of affect. *Cognition and Emotion*, 20(8), 1153–1169.

Cuddy, A. J., Rock, M. S., & Norton, M. I. (2007). Aid in the aftermath of Hurricane Katrina: Inferences of secondary emotions and intergroup helping. *Group Processes & Intergroup Relations*, 10(1), 107–118.

Cutright, K. M. (2011). The beauty of boundaries: When and why we seek structure in consumption. *Journal of Consumer Research*, 38(5), 775–790.

D'Argembeau, A., & Van der Linden, M. (2004). Phenomenal characteristics associated with projecting oneself back into the past and forward into the future: Influence of valence and temporal distance. *Consciousness and Cognition*, 13(4), 844–858.

Danovitch, J., & Bloom, P. (2009). Children's extension of disgust to physical and moral events. *Emotion*, 9(1), 107.

Davidoff, J. (2001). Language and perceptual categorisation. *Trends in Cognitive Sciences*, 5(9), 382–387.

Davis, S. N. (2003). Sex stereotypes in commercials targeted toward children: A content analysis. *Sociological Spectrum*, 23(4), 407–424.

DeBono, A., Shariff, A. F., Poole, S., & Muraven, M. (2017). Forgive us our trespasses: Priming a forgiving (but not a punishing) god increases unethical behavior. *Psychology of Religion and Spirituality*, 9(S1), S1.

Dehaene, S. (2011). *The number sense: How the mind creates mathematics.* New York, NY: Oxford University Press.

Deng, X., & Kahn, B. E. (2009). Is your product on the right side? The "location effect" on perceived product heaviness and package evaluation. *Journal of Marketing Research*, 46(6), 725–738.

Deng, X., Kahn, B. E., Unnava, H. R., & Lee, H. (2016). A "wide" variety: Effects of horizontal versus vertical display on assortment processing, perceived variety, and choice. *Journal of Marketing Research*, 53(5), 682–698.

Dewell, R. (2005). Dynamic patterns of CONTAINMENT. From Perception to Meaning: Image Schemas in Cognitive Linguistics, 369–394.

Dixson, A. F. (1983). Observations on the evolution and behavioral significance of "sexual skin" in female primates. In *Advances in the Study of Behavior* (Vol. 13, pp. 63–106). Academic Press.

DiYanni, C., & Kelemen, D. (2005). Time to get a new mountain? The role of function in children's conceptions of natural kinds. *Cognition*, 97(3), 327–335.

Dong, P., Huang, X., & Zhong, C. B. (2015). Ray of hope: Hopelessness increases preferences for brighter lighting. *Social Psychological and Personality Science*, 6(1), 84–91.

Dore, R. A., Smith, E. D., & Lillard, A. S. (2017). Children adopt the traits of characters in a narrative. *Child Development Research*, 2017.

Doyen, S., Klein, O., Pichon, C. L., & Cleeremans, A. (2012). Behavioral priming: it's all in the mind, but whose mind?. *PloS One*, 7(1), e29081.

Droit-Volet, S., Mermillod, M., Cocenas-Silva, R., & Gil, S. (2010). The effect of expectancy of a threatening event on time perception in human adults. *Emotion*, 10(6), 908.

Duduciuc, A. C. (2015). Advertising brands by means of sounds symbolism: the influence of vowels on perceived brand characteristics. *Studies and Scientific Researches. Economics Edition*, (21).

Durante, K. M., Griskevicius, V., Redden, J. P., & Edward White, A. (2015). Spending on daughters versus sons in economic recessions. *Journal of Consumer Research*, 42(3), 435–457.

Edgell, P., Gerteis, J., & Hartmann, D. (2006). Atheists as "other": Moral boundaries and cultural membership in American society. *American Sociological Review*, 71(2), 211–234.

Effron, D. A., Cameron, J. S., & Monin, B. (2009). Endorsing Obama licenses favoring whites. *Journal of Experimental Social Psychology*, 45(3), 590–593.

Ehrlich, K., & Johnson-Laird, P. N. (1982). Spatial descriptions and referential continuity. *Journal of Verbal Learning and Verbal Behavior*, 21(3), 296–306.

Ekman, P., & Friesen, W. V. (1971). Constants across cultures in the face and emotion. *Journal of Personality and Social Psychology*, 17(2), 124.

Elder, R. S., & Krishna, A. (2011). The "visual depiction effect" in advertising: Facilitating embodied mental simulation through product orientation. *Journal of Consumer Research*, 38(6), 988–1003.

Elder, R. S., Schlosser, A. E., Poor, M., & Xu, L. (2017). So close I can almost sense it: The interplay between sensory imagery and psychological distance. *Journal of Consumer Research*, 44(4), 877–894.

Elliot, A. J., & Aarts, H. (2011). Perception of the color red enhances the force and velocity of motor output. *Emotion*, 11(2), 445.

Elliot, A. J., & Niesta, D. (2008). Romantic red: red enhances men's attraction to women. *Journal of Personality and Social Psychology*, 95(5), 1150.

Elliot, A. J., & Pazda, A. D. (2012). Dressed for sex: Red as a female sexual signal in humans. *PLoS One*, 7(4), e34607.

Elliot, A. J., Greitemeyer, T., & Pazda, A. D. (2013). Women's use of red clothing as a sexual signal in intersexual interaction. *Journal of Experimental Social Psychology*, 49(3), 599–602.

Elliot, A. J., Niesta Kayser, D., Greitemeyer, T., Lichtenfeld, S., Gramzow, R. H., Maier, M. A., & Liu, H. (2010). Red, rank, and romance in women viewing men. *Journal of Experimental Psychology: General*, 139(3), 399.

Elliot, A. J., Tracy, J. L., Pazda, A. D., & Beall, A. T. (2013). Red enhances women's attractiveness to men: First evidence suggesting universality. *Journal of Experimental Social Psychology*, 49(1), 165–168.

Emery, N. J. (2000). The eyes have it: the neuroethology, function and evolution of social gaze. *Neuroscience & Biobehavioral Reviews*, 24(6), 581–604.

Englich, B., Mussweiler, T., & Strack, F. (2006). Playing dice with criminal sentences: The influence of irrelevant anchors on experts' judicial decision making. *Personality and Social Psychology Bulletin*, 32(2), 188–200.

Epley, N., Akalis, S., Waytz, A., & Cacioppo, J. T. (2008). Creating social connection through inferential reproduction: Loneliness and perceived agency in gadgets, gods, and greyhounds. *Psychological Science*, 19(2), 114–120.

Epley, N., Converse, B. A., Delbosc, A., Monteleone, G. A., & Cacioppo, J. T. (2009). Believers' estimates of God's beliefs are more egocentric than estimates of other people's beliefs. *Proceedings of the National Academy of Sciences*, 106(51), 21533–21538.

Epley, N., Keysar, B., Van Boven, L., & Gilovich, T. (2004). Perspective taking as egocentric anchoring and adjustment. *Journal of Personality and Social Psychology*, 87(3), 327.

Epley, N., Morewedge, C. K., & Keysar, B. (2004). Perspective taking in children and adults: Equivalent egocentrism but differential correction. *Journal of Experimental Social Psychology*, 40(6), 760–768.

Epley, N., Waytz, A., & Cacioppo, J. T. (2007). On seeing human: a three-factor theory of anthropomorphism. *Psychological Review*, 114(4), 864.

Escalas, J. E. (2004). Imagine yourself in the product: Mental simulation, narrative transportation, and persuasion. *Journal of Advertising*, 33(2), 37–48.

Eskine, K. J., Kacinik, N. A., & Prinz, J. J. (2011). A bad taste in the mouth: Gustatory disgust influences moral judgment. *Psychological Science*, 22(3), 295–299.

Esterman, M., & Yantis, S. (2009). Perceptual expectation evokes category-selective cortical activity. *Cerebral Cortex*, 20(5), 1245–1253.

Evans, L., & Davies, K. (2000). No sissy boys here: A content analysis of the representation of masculinity in elementary school reading textbooks. *Sex Roles*, 42(3–4), 255–270.

Face Facts (2019). Retrieved from: https://facefacts.scot/#average

Fajardo, T. M., & Townsend, C. (2016). Where you say it matters: Why packages are a more believable source of product claims than advertisements. *Journal of Consumer Psychology*, 26(3), 426–434.

Fanelli, D. (2009). How many scientists fabricate and falsify research? A systematic review and meta-analysis of survey data. *PloS One*, 4(5), e5738.

Fay, A. J., & Maner, J. K. (2012). Warmth, spatial proximity, and social attachment: The embodied perception of a social metaphor. *Journal of Experimental Social Psychology*, 48(6), 1369–1372.

Feingold, A. (1992). Good-looking people are not what we think. *Psychological Bulletin*, 111(2), 304.

Fessler, D. M., Eng, S. J., & Navarrete, C. D. (2005). Elevated disgust sensitivity in the first trimester of pregnancy: Evidence supporting the compensatory prophylaxis hypothesis. *Evolution and Human Behavior*, 26(4), 344–351.

Fetterman, A. K., Robinson, M. D., & Meier, B. P. (2012). Anger as "seeing red": Evidence for a perceptual association. *Cognition & Emotion*, 26(8), 1445–1458.

Filik, R., & Barber, E. (2011). Inner speech during silent reading reflects the reader's regional accent. *PloS One*, 6(10), e25782.

Fischer, M. H. (2001). Number processing induces spatial performance biases. *Neurology*, 57(5), 822–826.

Fisher, G., & Rangel, A. (2014). Symmetry in cold-to-hot and hot-to-cold valuation gaps. *Psychological Science*, 25(1), 120–127.

Fiske, S. T., Cuddy, A. J., & Glick, P. (2007). Universal dimensions of social cognition: Warmth and competence. *Trends in Cognitive Sciences*, 11(2), 77–83.

Fort, M., Martin, A., & Peperkamp, S. (2015). Consonants are more important than vowels in the bouba-kiki effect. *Language and Speech*, 58(2), 247–266.

Fox, C. R., & Rottenstreich, Y. (2003). Partition priming in judgment under uncertainty. *Psychological Science*, 14(3), 195–200.

Fox, C. R., Ratner, R. K., & Lieb, D. S. (2005). How subjective grouping of options influences choice and allocation: diversification bias and the phenomenon of partition dependence. *Journal of Experimental Psychology: General*, 134(4), 538.

Fraley, R. C., & Marks, M. J. (2010). Westermarck, Freud, and the incest taboo: does familial resemblance activate sexual attraction?. *Personality and Social Psychology Bulletin*, 36(9), 1202–1212.

Frank, M. G., & Gilovich, T. (1988). The dark side of self-and social perception: black uniforms and aggression in professional sports. *Journal of Personality and Social Psychology*, 54(1), 74.

Frank, C., Land, W. M., Popp, C., & Schack, T. (2014). Mental representation and mental practice: experimental investigation on the functional links between motor memory and motor imagery. *PloS One*, 9(4), e95175.

Friestad, M., & Wright, P. (1994). The persuasion knowledge model: How people cope with persuasion attempts. *Journal of Consumer Research*, 21(1), 1–31.

Fujita, K., Eyal, T., Chaiken, S., Trope, Y., & Liberman, N. (2008). Influencing attitudes toward near and distant objects. *Journal of Experimental Social Psychology*, 44(3), 562–572.

Gagl, B., Golch, J., Hawelka, S., Sassenhagen, J., Poeppel, D., & Fiebach, C. J. (2018). Reading at the speed of speech: the rate of eye movements aligns with auditory language processing. *bioRxiv*, 391896.

Galak, J., & Meyvis, T. (2011). The pain was greater if it will happen again: The effect of anticipated continuation on retrospective discomfort. *Journal of Experimental Psychology: General*, 140(1), 63.

Gendron, M., Roberson, D., van der Vyver, J. M., & Barrett, L. F. (2014a). Perceptions of emotion from facial expressions are not culturally universal: evidence from a remote culture. *Emotion*, 14(2), 251.

Gendron, M., Roberson, D., van der Vyver, J. M., & Barrett, L. F. (2014b). Cultural relativity in perceiving emotion from vocalizations. *Psychological Science*, 25(4), 911–920.

Giessner, S. R., & Schubert, T. W. (2007). High in the hierarchy: How vertical location and judgments of leaders' power are interrelated. *Organizational Behavior and Human Decision Processes*, 104(1), 30–44.

Gilovich, T., Medvec, V. H., & Savitsky, K. (2000). The spotlight effect in social judgment:

An egocentric bias in estimates of the salience of one's own actions and appearance. *Journal of Personality and Social Psychology*, 78(2), 211.
Glick, P., Zion, C., & Nelson, C. (1988). What mediates sex discrimination in hiring decisions? *Journal of Personality and Social Psychology*, 55(2), 178.
Goren, C. C., Sarty, M., & Wu, P. Y. (1975). Visual following and pattern discrimination of face-like stimuli by newborn infants. *Pediatrics*, 56(4), 544–549.
Gorn, G. J., Chattopadhyay, A., Sengupta, J., & Tripathi, S. (2004). Waiting for the web: how screen color affects time perception. *Journal of Marketing Research*, 41(2), 215–225.
Grammer, K., Fink, B., Møller, A. P., & Thornhill, R. (2003). Darwinian aesthetics: sexual selection and the biology of beauty. *Biological Reviews*, 78(3), 385–407.
Granqvist, P., & Hagekull, B. (2000). Religiosity, adult attachment, and why" singles" are more religious. The International *Journal for the Psychology of Religion*, 10(2), 111–123.
Griskevicius, V., Tybur, J. M., & Van den Bergh, B. (2010). Going green to be seen: status, reputation, and conspicuous conservation. *Journal of Personality and Social Psychology*, 98(3), 392.
Gross, J., Millett, A. L., Bartek, B., Bredell, K. H., & Winegard, B. (2014). Evidence for prosody in silent reading. *Reading Research Quarterly*, 49(2), 189–208.
Guéguen, N. (2012). Color and women hitchhikers' attractiveness: Gentlemen drivers prefer red. *Color Research & Application*, 37(1), 76–78.
Guéguen, N., & Jacob, C. (2014). Coffee cup color and evaluation of a beverage's "warmth quality". *Color Research & Application*, 39(1), 79–81.
Guido, G., Pichierri, M., Nataraajan, R., & Pino, G. (2016). Animated logos in mobile marketing communications: The roles of logo movement directions and trajectories. *Journal of Business Research*, 69(12), 6048–6057.
Hadjichristidis, C., Handley, S. J., Sloman, S. A., Evans, J. S. B., Over, D. E., & Stevenson, R. J. (2007). Iffy beliefs: Conditional thinking and belief change. *Memory & Cognition*, 35(8), 2052–2059.
Hagemann, N., Strauss, B., & Leißing, J. (2008). When the referee sees red.... *Psychological Science*, 19(8), 769–771.
Hagoort, P., Baggio, G., & Willems, R. M. (2009). Semantic unification. In *The Cognitive Neurosciences*, 4th ed. (pp. 819–836). MIT press.
Hagtvedt, H., & Brasel, S. A. (2016). Cross-modal communication: sound frequency influences consumer responses to color lightness. *Journal of Marketing Research*, 53(4), 551–562.
Haidt, J., Bjorklund, F., & Murphy, S. (2000). Moral dumbfounding: When intuition finds no reason. Unpublished manuscript, University of Virginia, 191–221.
Halim, M. L., Ruble, D. N., Tamis-LeMonda, C. S., Zosuls, K. M., Lurye, L. E., & Greulich, F. K. (2014). Pink frilly dresses and the avoidance of all things "girly": Children's appearance rigidity and cognitive theories of gender development. *Developmental Psychology*, 50(4), 1091.
Hanson-Vaux, G., Crisinel, A. S., & Spence, C. (2012). Smelling shapes: Crossmodal correspondences between odors and shapes. *Chemical Senses*, 38(2), 161–166.
Harrington, J., & Johnstone, A. (1987). The effects of equivalence classes on parsing phonemes into words in continuous speech recognition. *Computer Speech and Language*, 2, 273–288.
Harrison, M. A., & Hall, A. E. (2010). Anthropomorphism, empathy, and perceived communicative ability vary with phylogenetic relatedness to humans. *Journal of Social, Evolutionary, and Cultural Psychology*, 4(1), 34.
Hart, W., & Albarracín, D. (2011). Learning about what others were doing: Verb aspect

and attributions of mundane and criminal intent for past actions. *Psychological Science*, 22(2), 261–266.

Hartmann, M., & Mast, F. W. (2017). Loudness counts: Interactions between loudness, number magnitude, and space. *The Quarterly Journal of Experimental Psychology*, 70(7), 1305–1322.

Hassin, R. R. (2008). Being open minded without knowing why: Evidence from nonconscious goal pursuit. *Social Cognition*, 26(5), 578–592.

Havas, D. A., Glenberg, A. M., Gutowski, K. A., Lucarelli, M. J., & Davidson, R. J. (2010). Cosmetic use of botulinum toxin-A affects processing of emotional language. *Psychological Science*, 21(7), 895–900.

Hazeltine, R. E., Prinzmetal, W., & Elliott, K. (1997). If it's not there, where is it? Locating illusory conjunctions. *Journal of Experimental Psychology: Human Perception and Performance*, 23(1), 263.

Heinemann, A., Pfister, R., & Janczyk, M. (2013). Manipulating number generation: Loud+ long= large? *Consciousness and Cognition*, 22(4), 1332–1339.

Heiphetz, L., Lane, J. D., Waytz, A., & Young, L. L. (2016). How children and adults represent God's mind. *Cognitive Science*, 40(1), 121–144.

Henein, M. Y., Collaborators, G. R., Zhao, Y., Nicoll, R., Sun, L., Khir, A. W., . . . & Lindqvist, P. (2011). The human heart: application of the golden ratio and angle.

Herz, R. S. (2001). Ah, sweet skunk: why we like or dislike what we smell. *Cerebrum*, 3(4), 31–47.

Herz, R. S., & von Clef, J. (2001). The influence of verbal labeling on the perception of odors: evidence for olfactory illusions? *Perception*, 30(3), 381–391.

Herz, R. S., Beland, S. L., & Hellerstein, M. (2004). Changing odor hedonic perception through emotional associations in humans. *International Journal of Comparative Psychology*, 17(4).

Hill, R. A., & Barton, R. A. (2005). Red enhances human performance in contests. *Nature*, 435(7040), 293.

Hines, M. (2011). Gender development and the human brain. *Annual Review of Neuroscience*, 34, 69–88.

Ho, C. K. Y., Kuan, K. K., & Chau, P. Y. (2015). Temporal Features and Consumer Evaluations of Group-Buying: The Effects of Product Image Zooming. In *PACIS* (p. 122).

Holmes, K. J., & Lourenco, S. F. (2011). Common spatial organization of number and emotional expression: A mental magnitude line. *Brain and Cognition*, 77(2), 315–323.

Hong, J., & Sun, Y. (2011). Warm it up with love: The effect of physical coldness on liking of romance movies. *Journal of Consumer Research*, 39(2), 293–306.

Hope, A. L., & Jones, C. R. (2014). The impact of religious faith on attitudes to environmental issues and Carbon Capture and Storage (CCS) technologies: A mixed methods study. *Technology in Society*, 38, 48–59.

Horton, W. S., & Rapp, D. N. (2003). Out of sight, out of mind: Occlusion and the accessibility of information in narrative comprehension. *Psychonomic Bulletin & Review*, 10(1), 104–110.

Hsee, C. K., Yang, Y., Gu, Y., & Chen, J. (2008). Specification seeking: how product specifications influence consumer preference. *Journal of Consumer Research*, 35(6), 952–966.

Huang, L., & Galinsky, A. D. (2011). Mind–body dissonance: Conflict between the senses expands the mind's horizons. *Social Psychological and Personality Science*, 2(4), 351–359.

Huang, X. I., Zhang, M., Hui, M. K., & Wyer Jr, R. S. (2014). Warmth and conformity: The effects of ambient temperature on product preferences and financial decisions. *Journal of Consumer Psychology*, 24(2), 241–250.

Hubbard, T. L. (2005). Representational momentum and related displacements in spatial memory: A review of the findings. *Psychonomic Bulletin & Review*, 12(5), 822–851.

Hyde, J. S. (2005). The gender similarities hypothesis. *American Psychologist*, 60(6), 581.
Imschloss, M., & Kuehnl, C. (2017). Don't ignore the floor: Exploring multisensory atmospheric congruence between music and flooring in a retail environment. *Psychology & Marketing*, 34(10), 931–945.
Inbar, Y., Pizarro, D. A., & Bloom, P. (2009). Conservatives are more easily disgusted than liberals. *Cognition and Emotion*, 23(4), 714–725.
Inbar, Y., Pizarro, D. A., & Bloom, P. (2012). Disgusting smells cause decreased liking of gay men. *Emotion*, 12(1), 23.
Isaac, M. S., & Schindler, R. M. (2013). The top-ten effect: Consumers' subjective categorization of ranked lists. *Journal of Consumer Research*, 40(6), 1181–1202.
Jablonski, N. G. (2004). The evolution of human skin and skin color. *Annual Review Anthropology*, 33, 585–623.
Jablonski, N. G., & Chaplin, G. (2000). The evolution of human skin coloration. *Journal of Human Evolution*, 39(1), 57–106.
Jackson, L. A., Hunter, J. E., & Hodge, C. N. (1995). Physical attractiveness and intellectual competence: A meta-analytic review. *Social Psychology Quarterly*, 58, 108–108.
Jenkins, D., & Johnston, L. B. (2004). Unethical treatment of gay and lesbian people with conversion therapy. *Families in Society*, 85(4), 557–561.
Jensen, G. F. (2006). Religious cosmologies and homicide rates among nations: A closer look.
Johnson, D. D. (2005). God's punishment and public goods human. *Nature*, 16(4), 410–446.
Johnson, D., & Krüger, O. (2004). The good of wrath: Supernatural punishment and the evolution of cooperation. *Political Theology*, 5(2), 159–176.
Johnson, E., Johnson, N. B., & Shanthikumar, D. (2007). Round numbers and security returns. Unpublished Working Paper.
Johnson, M. K., Rowatt, W. C., & LaBouff, J. (2010). Priming Christian religious concepts increases racial prejudice. *Social Psychological and Personality Science*, 1(2), 119–126.
Johnson, M. K., Rowatt, W. C., & LaBouff, J. P. (2012). Religiosity and prejudice revisited: In-group favoritism, out-group derogation, or both? *Psychology of Religion and Spirituality*, 4(2), 154.
Johnson, S. K., Podratz, K. E., Dipboye, R. L., & Gibbons, E. (2010). Physical attractiveness biases in ratings of employment suitability: Tracking down the "beauty is beastly" effect. The *Journal of Social Psychology*, 150(3), 301–318.
Jostmann, N. B., Lakens, D., & Schubert, T. W. (2009). Weight as an embodiment of importance. *Psychological Science,* 20(9), 1169–1174.
Judd, C. M., James-Hawkins, L., Yzerbyt, V., & Kashima, Y. (2005). Fundamental dimensions of social judgment: understanding the relations between judgments of competence and warmth. *Journal of Personality and Social Psychology*, 89(6), 899.
Jung, K., Shavitt, S., Viswanathan, M., & Hilbe, J. M. (2014). Female hurricanes are deadlier than male hurricanes. *Proceedings of the National Academy of Sciences*, 111(24), 8782–8787.
Kahneman, D. (2011). Thinking, fast and slow. Macmillan.
Kahneman, D., Treisman, A., & Gibbs, B. J. (1992). The reviewing of object files: Object-specific integration of information. *Cognitive Psychology*, 24(2), 175–219.
Kamalski, J. (2007). Coherence marking, comprehension and persuasion. On the processing and representation of discourse, 158.
Kareklas, I., Brunel, F. F., & Coulter, R. A. (2014). Judgment is not color blind: The impact of automatic color preference on product and advertising preferences. *Journal of Consumer Psychology*, 24(1), 87–95.
Karmarkar, U. R., & Bollinger, B. (2015). BYOB: How bringing your own shopping bags leads to treating yourself and the environment. *Journal of Marketing*, 79(4), 1–15.

Kaspar, K. (2013a). A weighty matter: Heaviness influences the evaluation of disease severity, drug effectiveness, and side effects. *PLoS One*, 8(11), e78307.

Kaspar, K. (2013b). Washing one's hands after failure enhances optimism but hampers future performance. *Social Psychological and Personality Science*, 4(1), 69–73.

Kaufman, G. F., & Libby, L. K. (2012). Changing beliefs and behavior through experience-taking. *Journal of Personality and Social Psychology*, 103(1), 1.

Kay, A. C., Gaucher, D., McGregor, I., & Nash, K. (2010). Religious belief as compensatory control. *Personality and Social Psychology Review*, 14(1), 37–48.

Kay, A. C., Gaucher, D., Napier, J. L., Callan, M. J., & Laurin, K. (2008). God and the government: testing a compensatory control mechanism for the support of external systems. *Journal of Personality and Social Psychology*, 95(1), 18.

Kay, A. C., Shepherd, S., Blatz, C. W., Chua, S. N., & Galinsky, A. D. (2010). For God (or) country: the hydraulic relation between government instability and belief in religious sources of control. *Journal of Personality and Social Psychology*, 99(5), 725.

Kelemen, D., & Rosset, E. (2009). The human function compunction: Teleological explanation in adults. *Cognition*, 111(1), 138–143.

Kellaris, J. J., & Rice, R. C. (1993). The influence of tempo, loudness, and gender of listener on responses to music. *Psychology & Marketing*, 10(1), 15–29.

Kelly, M. H., & Bock, J. K. (1988). Stress in time. *Journal of Experimental Psychology: human perception* and performance, 14(3), 389.

Kennison, S. M., & Trofe, J. L. (2003). Comprehending pronouns: A role for word-specific gender stereotype information. *Journal of Psycholinguistic Research*, 32(3), 355–378.

Kenrick, D. T., & Gutierres, S. E. (1980). Contrast effects and judgments of physical attractiveness: When beauty becomes a social problem. *Journal of Personality and Social Psychology*, 38(1), 131.

Khan, U., & Dhar, R. (2006). Licensing effect in consumer choice. *Journal of Marketing Research*, 43(2), 259–266.

Kille, D. R., Forest, A. L., & Wood, J. V. (2013). Tall, dark, and stable: Embodiment motivates mate selection preferences. *Psychological Science*, 24(1), 112–114.

King, D., & Janiszewski, C. (2011a). The sources and consequences of the fluent processing of numbers. *Journal of Marketing Research*, 48(2), 327–341.

King, D., & Janiszewski, C. (2011b). Affect-gating. *Journal of Consumer Research*, 38(4), 697–711.

Kiper, J., & Meier, J. (2015). The problems and origins of belief in Big Gods. *Religion, Brain & Behavior*, 5(4), 298–305.

Kirkpatrick, L. A., Shillito, D. J., & Kellas, S. L. (1999). Loneliness, social support, and perceived relationships with God. *Journal of Social and Personal Relationships*, 16(4), 513–522.

Klink, R. R. (2000). Creating brand names with meaning: The use of sound symbolism. *Marketing Letters*, 11(1), 5–20.

Kniffin, K. M., & Shimizu, M. (2016). Sounds that make you smile and share: a phonetic key to prosociality and engagement. *Marketing Letters*, 27(2), 273–283.

Kohut, A., Wike, R., Bell, J., Horowitz, J. M., Simmons, K., Stokes, B., . . . & Devlin, K. (2013). The global divide on homosexuality. *Pew Research Center*, 4.

Kok, P., Failing, M. F., & de Lange, F. P. (2014). Prior expectations evoke stimulus templates in the primary visual cortex. *Journal of Cognitive Neuroscience*, 26(7), 1546–1554.

Koo, M., & Fishbach, A. (2012). The small-area hypothesis: Effects of progress monitoring on goal adherence. *Journal of Consumer Research*, 39(3), 493–509.

Krishna, A., & Morrin, M. (2007). Does touch affect taste? The perceptual transfer of product container haptic cues. *Journal of Consumer Research*, 34(6), 807–818.

Krishna, A., Elder, R. S., & Caldara, C. (2010). Feminine to smell but masculine to touch? Multisensory congruence and its effect on the aesthetic experience. *Journal of Consumer Psychology*, 20(4), 410–418.

Kwon, M., & Adaval, R. (2017). Going against the Flow: The Effects of Dynamic Sensorimotor Experiences on Consumer Choice. *Journal of Consumer Research*, 44(6), 1358–1378.

Labrecque, L. I. (2010). The marketer's prismatic palette: Essays on the importance of color in marketing with implications for brand personality.

Laeng, B., Vermeer, O., & Sulutvedt, U. (2013). Is beauty in the face of the beholder?. *PloS One*, 8(7), e68395.

Laham, S. M., Koval, P., & Alter, A. L. (2012). The name-pronunciation effect: Why people like Mr. Smith more than Mr. Colquhoun. *Journal of Experimental Social Psychology*, 48(3), 752–756.

Lakens, D., Semin, G. R., & Foroni, F. (2011). Why your highness needs the people. *Social Psychology*.

Lakoff, G. (2008) The Neural Theory of Metaphor. *The Cambridge Handbook of Metaphor and Thought*, (pp. 17–38). New York, NY: Cambridge University Press.

Lakoff, G., & Johnson, M. (1980). Metaphors we live by. Chicago: University of Chicago Press.

Lakoff, G., & Johnson, M. (1999). Philosophy in the flesh (Vol. 4). New York, NY: Basic Books.

Langlois, J. H., & Roggman, L. A. (1990). Attractive faces are only average. *Psychological Science*, 1(2), 115–121.

Langton, S. R., Watt, R. J., & Bruce, V. (2000). Do the eyes have it? Cues to the direction of social attention. *Trends in Cognitive Sciences*, 4(2), 50–59.

Larson, C. L., Aronoff, J., & Stearns, J. J. (2007). The shape of threat: Simple geometric forms evoke rapid and sustained capture of attention. *Emotion*, 7(3), 526.

Lassiter, G. D., Munhall, P. J., Berger, I. P., Weiland, P. E., Handley, I. M., & Geers, A. L. (2005). Attributional complexity and the camera perspective bias in videotaped confessions. *Basic and Applied Social Psychology*, 27(1), 27–35.

Lassiter, G., Diamond, S. S., Schmidt, H. C., & Elek, J. K. (2007). Evaluating videotaped confessions: Expertise provides no defense against the camera-perspective effect. *Psychological Science*, 18(3), 224–226.

Laurin, K., Schumann, K., & Holmes, J. G. (2014). A relationship with God? Connecting with the divine to assuage fears of interpersonal rejection. *Social Psychological and Personality Science*, 5(7), 777–785.

Lederman, S. J., & Klatzky, R. L. (1987). Hand movements: A window into haptic object recognition. *Cognitive Psychology*, 19(3), 342–368.

Lee, A., & Ji, L. J. (2014). Moving away from a bad past and toward a good future: Feelings influence the metaphorical understanding of time. *Journal of Experimental Psychology: General*, 143(1), 21.

Lee, S. W., & Schwarz, N. (2010a). Dirty hands and dirty mouths: Embodiment of the moral-purity metaphor is specific to the motor modality involved in moral transgression. *Psychological Science*, 21(10), 1423–1425.

Lee, S. W., & Schwarz, N. (2010b). Washing away postdecisional dissonance. *Science*, 328(5979), 709–709.

Lee, S. W., & Schwarz, N. (2012). Bidirectionality, mediation, and moderation of metaphorical effects: the embodiment of social suspicion and fishy smells. *Journal of Personality and Social Psychology*, 103(5), 737.

Lee, H., Fujita, K., Deng, X., & Unnava, H. R. (2016). The role of temporal distance on the color of future-directed imagery: A construal-level perspective. *Journal of Consumer Research*, 43(5), 707–725.

Lee, S. H. M., Rotman, J. D., & Perkins, A. W. (2014). Embodied cognition and social consumption: Self-regulating temperature through social products and behaviors. *Journal of Consumer Psychology*, 24(2), 234–240.

Leonhardt, J. M., Catlin, J. R., & Pirouz, D. M. (2015). Is your product facing the ad's center? facing direction affects processing fluency and ad evaluation. *Journal of Advertising*, 44(4), 315–325.

Lessard, D. A., Linkenauger, S. A., & Proffitt, D. R. (2009). Look before you leap: Jumping ability affects distance perception. *Perception*, 38(12), 1863–1866.

Levav, J., & Zhu, R. (2009). Seeking freedom through variety. *Journal of Consumer Research*, 36(4), 600–610.

Levine, M. (2009, March). Share my ride. The New York Times. Retrieved from https://www.nytimes.com/2009/03/08/magazine/08Zipcar-t.html

Levine, M., Prosser, A., Evans, D., & Reicher, S. (2005). Identity and emergency intervention: How social group membership and inclusiveness of group boundaries shape helping behavior. *Personality and Social Psychology Bulletin*, 31(4), 443–453.

Li, X., Wei, L., & Soman, D. (2010). Sealing the emotions genie: The effects of physical enclosure on psychological closure. *Psychological Science*, 21(8), 1047–1050.

Liberman, A. M., & Mattingly, I. G. (1985). The motor theory of speech perception revised. *Cognition*, 21(1), 1–36.

Liberman, N., & Trope, Y. (1998). The role of feasibility and desirability considerations in near and distant future decisions: A test of temporal construal theory. *Journal of Personality and Social Psychology*, 75(1), 5.

Lightdale, J. R., & Prentice, D. A. (1994). Rethinking sex differences in aggression: Aggressive behavior in the absence of social roles. *Personality and Social Psychology Bulletin*, 20(1), 34–44.

List, A., Iordanescu, L., Grabowecky, M., & Suzuki, S. (2014). Haptic guidance of overt visual attention. *Attention, Perception, & Psychophysics*, 76(8), 2221–2228.

Lockwood, G., & Dingemanse, M. (2015). Iconicity in the lab: A review of behavioral, developmental, and neuroimaging research into sound-symbolism. *Frontiers in Psychology*, 6, 1246.

Loetscher, T., Bockisch, C. J., Nicholls, M. E., & Brugger, P. (2010). Eye position predicts what number you have in mind. *Current Biology*, 20(6), R264–R265.

Loewenstein, G., Nagin, D., & Paternoster, R. (1997). The effect of sexual arousal on expectations of sexual forcefulness. *Journal of Research in Crime and Delinquency*, 34(4), 443–473.

Lombrozo, T., Kelemen, D., & Zaitchik, D. (2007). Inferring design: Evidence of a preference for teleological explanations in patients with Alzheimer's disease. *Psychological Science*, 18(11), 999–1006.

Lowrey, T. M., & Shrum, L. J. (2007). Phonetic symbolism and brand name preference. *Journal of Consumer Research*, 34(3), 406–414.

Lundqvist, D., Esteves, F., & Ohman, A. (1999). The face of wrath: Critical features for conveying facial threat. *Cognition & Emotion*, 13(6), 691–711.

Lurye, L. E., Zosuls, K. M., & Ruble, D. N. (2008). Gender identity and adjustment:

Maass, A., Suitner, C., Favaretto, X., & Cignacchi, M. (2009). Groups in space: Stereotypes and the spatial agency bias. *Journal of Experimental Social Psychology*, 45(3), 496–504.

Macrae, C. N., Alnwick, K. A., Milne, A. B., & Schloerscheidt, A. M. (2002). Person perception across the menstrual cycle: Hormonal influences on social-cognitive functioning. *Psychological Science*, 13(6), 532–536.

Madden, C. J., & Zwaan, R. A. (2003). How does verb aspect constrain event representations? *Memory & Cognition*, 31(5), 663–672.

Magid, M., Finzi, E., Kruger, T. H. C., Robertson, H. T., Keeling, B. H., Jung, S., . . . &

Wollmer, M. A. (2015). Treating depression with botulinum toxin: a pooled analysis of randomized controlled trials. *Pharmacopsychiatry*, 25(06), 205–210.
Maglio, S. J., & Polman, E. (2014). Spatial orientation shrinks and expands psychological distance. *Psychological Science*, 25(7), 1345–1352.
Maglio, S. J., & Polman, E. (2016). Revising probability estimates: Why increasing likelihood means increasing impact. *Journal of Personality and Social Psychology*, 111(2), 141.
Maglio, S. J., Rabaglia, C. D., Feder, M. A., Krehm, M., & Trope, Y. (2014). Vowel sounds in words affect mental construal and shift preferences for targets. *Journal of Experimental Psychology: General*, 143(3), 1082.
Magnus, M. (2001). What's in a Word? Studies in Phonosemantics. Det historisk-filosofiske fakultet.
Malhotra, D. K. (2008). (When) are Religious People Nicer? Religious Salience and the 'Sunday Effect' on Pro-Social Behavior.
Mandler, J. M. (1988). How to build a baby: On the development of an accessible representational system. *Cognitive Development*, 3(2), 113–136.
Mandler, J. M. (1992). How to build a baby: II. Conceptual primitives. *Psychological Review*, 99(4), 587.
Mandler, J. M. (2004). The foundations of mind: Origins of conceptual thought. Oxford University Press.
Mann, N. H., & Kawakami, K. (2012). The long, steep path to equality: Progressing on egalitarian goals. *Journal of Experimental Psychology: General*, 141(1), 187.
Mar, R. A., Oatley, K., & Peterson, J. B. (2009). Exploring the link between reading fiction and empathy: Ruling out individual differences and examining outcomes. *Communications*, 34(4), 407–428.
Martin, C. L., & Ruble, D. (2004). Children's search for gender cues: Cognitive perspectives on gender development. *Current Directions in Psychological Science*, 13(2), 67–70.
Matlock, T. (2011). The conceptual motivation of aspect. Motivation in Grammar and the Lexicon, 27, 133.
Mazar, N., & Zhong, C. B. (2010). Do green products make us better people? *Psychological Science*, 21(4), 494–498.
McCann, S. J. (1999). Threatening times and fluctuations in American church memberships. *Personality and Social Psychology Bulletin*, 25(3), 325–336.
McCrink, K., & Wynn, K. (2009). Operational momentum in large-number addition and subtraction by 9-month-olds. *Journal of Experimental Child Psychology*, 103(4), 400–408.
McGlone, M. S., & Tofighbakhsh, J. (2000). Birds of a feather flock conjointly (?): Rhyme as reason in aphorisms. *Psychological Science*, 11(5), 424–428.
McIntosh, D. N., Silver, R. C., & Wortman, C. B. (1993). Religion's role in adjustment to a negative life event: coping with the loss of a child. *Journal of Personality and Social Psychology*, 65(4), 812.
Meert, K., Pandelaere, M., & Patrick, V. M. (2014). Taking a shine to it: How the preference for glossy stems from an innate need for water. *Journal of Consumer Psychology*, 24(2), 195–206.
Mehta, R., & Zhu, R. J. (2009). Blue or red? Exploring the effect of color on cognitive task performances. *Science*, 323(5918), 1226–1229.
Meier, B. P., & Dionne, S. (2009). Downright sexy: Verticality, implicit power, and perceived physical attractiveness. *Social Cognition*, 27(6), 883–892.
Meier, B. P., & Robinson, M. D. (2004). Why the sunny side is up: Associations between affect and vertical position. *Psychological Science*, 15(4), 243–247.
Meier, B. P., & Robinson, M. D. (2006). Does "feeling down" mean seeing down? Depressive symptoms and vertical selective attention. *Journal of Research in Personality*, 40(4), 451–461.

Meier, B. P., Fetterman, A. K., & Robinson, M. D. (2015). Black and white as valence cues. *Social Psychology*.

Meier, B. P., Fetterman, A. K., Robinson, M. D., & Lappas, C. M. (2015). The myth of the angry atheist. *The Journal of Psychology*, 149(3), 219–238.

Meier, B. P., Moller, A. C., Chen, J. J., & Riemer-Peltz, M. (2011). Spatial metaphor and real estate: North–South location biases housing preference. *Social Psychological and Personality Science*, 2(5), 547–553.

Meier, B. P., Robinson, M. D., & Clore, G. L. (2004). Why good guys wear white: Automatic inferences about stimulus valence based on brightness. *Psychological Science*, 15(2), 82–87.

Meissner, C. A., & Brigham, J. C. (2001). Thirty years of investigating the own-race bias in memory for faces: A meta-analytic review. *Psychology, Public Policy, and Law*, 7(1), 3.

Meister, I. G., Krings, T., Foltys, H., Boroojerdi, B., Müller, M., Töpper, R., & Thron, A. (2004). Playing piano in the mind—an fMRI study on music imagery and performance in pianists. *Cognitive Brain Research*, 19(3), 219–228.

Melamed, L., & Moss, M. K. (1975). The effect of context on ratings of attractiveness of photographs. *The Journal of Psychology*, 90(1), 129–136.

Meyers-Levy, J., & Zhu, R. (2007). The influence of ceiling height: The effect of priming on the type of processing that people use. *Journal of Consumer Research*, 34(2), 174–186.

Miles, L., Nind, L., & Macrae, C. (2010). Moving through time. *Psychological Science*, 21(2), 222.

Milkman, K. L., Akinola, M., & Chugh, D. (2012). Temporal distance and discrimination: An audit study in academia. *Psychological Science*, 23(7), 710–717.

Mishra, A. (2008). Influence of contagious versus noncontagious product groupings on consumer preferences. *Journal of Consumer Research*, 36(1), 73–82.

Mishra, A., & Mishra, H. (2010). Border bias: The belief that state borders can protect against disasters. *Psychological Science*, 21(11), 1582–1586.

Moeller, S. K., Robinson, M. D., & Zabelina, D. L. (2008). Personality dominance and preferential use of the vertical dimension of space: Evidence from spatial attention paradigms. *Psychological Science*, 19(4), 355–361.

Monaghan, P., Christiansen, M. H., & Fitneva, S. A. (2011). The arbitrariness of the sign: Learning advantages from the structure of the vocabulary. *Journal of Experimental Psychology: General*, 140(3), 325.

Mondloch, C. J., Le Grand, R., & Maurer, D. (2003). Early visual experience is necessary for the development of some—but not all—aspects of face processing. *The development of face processing in infancy and early childhood: Current perspectives*, 99–117.

Morewedge, C. K., & Buechel, E. C. (2013). Motivated underpinnings of the impact bias in affective forecasts. *Emotion*, 13(6), 1023.

Morewedge, C. K., Huh, Y. E., & Vosgerau, J. (2010). Thought for food: Imagined consumption reduces actual consumption. *Science*, 330(6010), 1530–1533.

Moss-Racusin, C. A., Dovidio, J. F., Brescoll, V. L., Graham, M. J., & Handelsman, J. (2012). Science faculty's subtle gender biases favor male students. *Proceedings of the National Academy of Sciences*, 109(41), 16474–16479.

Mourey, J. A., Olson, J. G., & Yoon, C. (2017). Products as pals: Engaging with anthropomorphic products mitigates the effects of social exclusion. *Journal of Consumer Research*, 44(2), 414–431.

Mulatti, C., Treccani, B., & Job, R. (2014). The role of the sound of objects in object identification: evidence from picture naming. *Frontiers in Psychology*, 5, 1139.

Nelson, L. D., & Simmons, J. P. (2009). On southbound ease and northbound fees: Literal consequences of the metaphoric link between vertical position and cardinal direction. *Journal of Marketing Research*, 46(6), 715–724.

Neuhoff, J. G. (2001). An adaptive bias in the perception of looming auditory motion. *Ecological Psychology*, 13(2), 87–110.

Nicholls, M. E., Clode, D., Wood, S. J., & Wood, A. G. (1999). Laterality of expression in portraiture: Putting your best cheek forward. *Proceedings of the Royal Society of London. Series B: Biological Sciences*, 266(1428), 1517–1522.

Nickerson, R. S. (1999). How we know—and sometimes misjudge—what others know: Imputing one's own knowledge to others. *Psychological Bulletin*, 125(6), 737.

Norenzayan, A. (2013). *Big gods: How religion transformed cooperation and conflict.* Princeton University Press.

Norenzayan, A. (2014). Does religion make people moral? *Behaviour*, 151(2–3), 365–384.

Norenzayan, A., & Hansen, I. G. (2006). Belief in supernatural agents in the face of death. *Personality and Social Psychology Bulletin*, 32(2), 174–187.

Norenzayan, A., Gervais, W. M., & Trzesniewski, K. H. (2012). Mentalizing deficits constrain belief in a personal God. *PloS One*, 7(5), e36880.

Norman, D. (2013). *The design of everyday things: Revised and expanded edition.* New York, NY: Basic books.

Nunes, J. C., & Boatwright, P. (2004). Incidental prices and their effect on willingness to pay. *Journal of Marketing Research*, 41(4), 457–466.

Nunes, J. C., Ordanini, A., & Valsesia, F. (2015). The power of repetition: repetitive lyrics in a song increase processing fluency and drive market success. *Journal of Consumer Psychology*, 25(2), 187–199.

Ohala, J. J., Hinton, L., & Nichols, J. (1997, August). Sound symbolism. In *Proc. 4th Seoul International Conference on Linguistics* [SICOL] (pp. 98–103).

Ohtsuka, K., & Brewer, W. F. (1992). Discourse organization in the comprehension of temporal order in narrative texts. *Discourse Processes*, 15(3), 317–336.

Oliveri, M., Vicario, C. M., Salerno, S., Koch, G., Turriziani, P., Mangano, R., . . . & Caltagirone, C. (2008). Perceiving numbers alters time perception. *Neuroscience Letters*, 438(3), 308–311.

Omotehinwa, T. O., & Ramon, S. O. (2013). Fibonacci numbers and Golden ratio in mathematics and science. *International Journal of Computer and Information Technology* (ISSN: 2279-0764) Volume.

Open Science Collaboration. (2015). Estimating the reproducibility of psychological science. *Science*, 349(6251), aac4716.

Oppenheimer, D. M., LeBoeuf, R. A., & Brewer, N. T. (2008). Anchors aweigh: A demonstration of cross-modality anchoring and magnitude priming. *Cognition*, 106(1), 13–26.

Ozturk, O., Krehm, M., & Vouloumanos, A. (2013). Sound symbolism in infancy: evidence for sound–shape cross-modal correspondences in 4-month-olds. *Journal of Experimental Child Psychology*, 114(2), 173–186.

Palmer, S. (1981). Canonical perspective and the perception of objects. *Attention and Performance*, 135–151.

Palmer, S. E., & Schloss, K. B. (2010). An ecological valence theory of human color preference. *Proceedings of the National Academy of Sciences*, 107(19), 8877–8882.

Palmer, S. E., Gardner, J. S., & Wickens, T. D. (2008). Aesthetic issues in spatial composition: Effects of position and direction on framing single objects. *Spatial Vision*, 21(3), 421–450.

Palumbo, L., Ruta, N., & Bertamini, M. (2015). Comparing angular and curved shapes in terms of implicit associations and approach/avoidance responses. *PloS One*, 10(10), e0140043.

Pandelaere, M., Briers, B., & Lembregts, C. (2011). How to make a 29% increase look bigger: The unit effect in option comparisons. *Journal of Consumer Research*, 38(2), 308–322.

Parsons, L. M. (1987). Imagined spatial transformation of one's body. *Journal of Experimental Psychology: General*, 116(2), 172.

Pascalis, O., Scott, L. S., Kelly, D. J., Shannon, R. W., Nicholson, E., Coleman, M., & Nelson, C. A. (2005). Plasticity of face processing in infancy. *Proceedings of the national academy of sciences*, 102(14), 5297–5300.

Pasterski, V., Hindmarsh, P., Geffner, M., Brook, C., Brain, C., & Hines, M. (2007). Increased aggression and activity level in 3-to 11-year-old girls with congenital adrenal hyperplasia (CAH). *Hormones and Behavior*, 52(3), 368–374.

Pazda, A. D., Elliot, A. J., & Greitemeyer, T. (2012). Sexy red: Perceived sexual receptivity mediates the red-attraction relation in men viewing woman. *Journal of Experimental Social Psychology*, 48(3), 787–790.

Pazzaglia, M., Pizzamiglio, L., Pes, E., & Aglioti, S. M. (2008). The sound of actions in apraxia. *Current Biology*, 18(22), 1766–1772.

Pecher, D., Van Dantzig, S., Boot, I., Zanolie, K., & Huber, D. E. (2010). Congruency between word position and meaning is caused by task-induced spatial attention. *Frontiers in Psychology*, 1, 30.

Peck, T. C., Seinfeld, S., Aglioti, S. M., & Slater, M. (2013). Putting yourself in the skin of a black avatar reduces implicit racial bias. *Consciousness and Cognition*, 22(3), 779–787.

Pennington, N., & Hastie, R. (1986). Evidence evaluation in complex decision making. *Journal of Personality and Social Psychology*, 51(2), 242.

Penton-Voak, I. S., & Perrett, D. I. (2000). Female preference for male faces changes cyclically: Further evidence. *Evolution and Human Behavior*, 21(1), 39–48.

Peoples, H. C., & Marlowe, F. W. (2012). Subsistence and the evolution of religion. *Human Nature*, 23(3), 253–269.

Perrett, D., Penton-Voak, I. S., Little, A. C., Tiddeman, B. P., Burt, D. M., Schmidt, N., . . . & Barrett, L. (2002). Facial attractiveness judgements reflect learning of parental age characteristics. *Proceedings of the Royal Society of London. Series B: Biological Sciences*, 269(1494), 873–880.

Phillips, W., & Boroditsky, L. (2003). Can quirks of grammar affect the way you think? Grammatical gender and object concepts. In *Proceedings of the Annual Meeting of the Cognitive Science Society* (Vol. 25, No. 25).

Piazza, J., Bering, J. M., & Ingram, G. (2011). "Princess Alice is watching you": Children's belief in an invisible person inhibits cheating. *Journal of Experimental Child Psychology*, 109(3), 311–320.

Piqueras-Fiszman, B., & Spence, C. (2012). The weight of the container influences expected satiety, perceived density, and subsequent expected fullness. *Appetite*, 58(2), 559–562.

Pocheptsova, A., Labroo, A. A., & Dhar, R. (2010). Making products feel special: When metacognitive difficulty enhances evaluation. *Journal of Marketing Research*, 47(6), 1059–1069.

Pollak, S. D., & Sinha, P. (2002). Effects of early experience on children's recognition of facial displays of emotion. *Developmental Psychology*, 38(5), 784.

Pope, D., & Simonsohn, U. (2011). Round numbers as goals: Evidence from baseball, SAT takers, and the lab. *Psychological Science*, 22(1), 71–79.

Powers, K. E., Worsham, A. L., Freeman, J. B., Wheatley, T., & Heatherton, T. F. (2014). Social connection modulates perceptions of animacy. *Psychological Science*, 25(10), 1943–1948.

Preston, J. L., & Ritter, R. S. (2013). Different effects of religion and God on prosociality with the ingroup and outgroup. *Personality and Social Psychology Bulletin*, 39(11), 1471–1483.

Preston, J. L., Ritter, R. S., & Ivan Hernandez, J. (2010). Principles of religious prosociality: A review and reformulation. *Social and Personality Psychology Compass*, 4(8), 574–590.

Principe, C. P., & Langlois, J. H. (2012). Shifting the prototype: Experience with faces influences affective and attractiveness preferences. *Social Cognition*, 30(1), 109–120.

Puccinelli, N., Tickle-Degnen, L., & Rosenthal, R. (2006). Stage Left, Stage Right? Position Effects on Perception of Spokesperson. *ACR North American Advances*.

Purzycki, B. G., Apicella, C., Atkinson, Q. D., Cohen, E., McNamara, R. A., Willard, A. K., . . . & Henrich, J. (2016). Moralistic gods, supernatural punishment and the expansion of human sociality. *Nature*, 530(7590), 327.

Pylyshyn, Z. (1989). The role of location indexes in spatial perception: A sketch of the FINST spatial-index model. *Cognition*, 32(1), 65–97.

Rabelo, A. L., Keller, V. N., Pilati, R., & Wicherts, J. M. (2015). No effect of weight on judgments of importance in the moral domain and evidence of publication bias from a meta-analysis. *PloS One*, 10(8), e0134808.

Radvansky, G. A., Zwaan, R. A., Federico, T., & Franklin, N. (1998). Retrieval from temporally organized situation models. *Journal of Experimental Psychology: Learning, Memory, and Cognition*, 24(5), 1224.

Raghubir, P., & Srivastava, J. (2002). Effect of face value on product valuation in foreign currencies. *Journal of Consumer Research*, 29(3), 335–347.

Raghubir, P., & Valenzuela, A. (2006). Center-of-inattention: Position biases in decision-making. *Organizational Behavior and Human Decision Processes*, 99(1), 66–80.

Raghubir, P., Morwitz, V. G., & Chakravarti, A. (2011). Spatial categorization and time perception: Why does it take less time to get home? *Journal of Consumer Psychology*, 21(2), 192–198.

Ramsey-Rennels, J. L., & Langlois, J. H. (2006). Infants' differential processing of female and male faces. *Current Directions in Psychological Science*, 15(2), 59–62.

Rapp, D. N., & Taylor, H. A. (2004). Interactive dimensions in the construction of mental representations for text. *Journal of Experimental Psychology: Learning, Memory, and Cognition*, 30(5), 988.

Reber, R., & Schwarz, N. (1999). Effects of perceptual fluency on judgments of truth. *Consciousness and Cognition*, 8(3), 338–342.

Rees, T. J. (2009). Is personal insecurity a cause of cross-national differences in the intensity of religious belief?

Rennels, J. L., & Davis, R. E. (2008). Facial experience during the first year. *Infant Behavior and Development*, 31(4), 665–678.

Rensink, R. A. (2000). The dynamic representation of scenes. *Visual Cognition*, 7(1–3), 17–42.

Rentfrow, P. J., Goldberg, L. R., Stillwell, D. J., Kosinski, M., Gosling, S. D., & Levitin, D. J. (2012). The song remains the same: A replication and extension of the MUSIC model. *Music Perception: An Interdisciplinary Journal*, 30(2), 161–185.

Reynolds, D. J., Garnham, A., & Oakhill, J. (2006). Evidence of immediate activation of gender information from a social role name. *The Quarterly Journal of Experimental Psychology*, 59(05), 886–903.

Rim, S., Hansen, J., & Trope, Y. (2013). What happens why? Psychological distance and focusing on causes versus consequences of events. *Journal of Personality and Social Psychology*, 104(3), 457.

Robinson, M. D., Cassidy, D. M., Boyd, R. L., & Fetterman, A. K. (2015). The politics of time: Conservatives differentially reference the past and liberals differentially reference the future. *Journal of Applied Social Psychology*, 45(7), 391–399.

Romero, M., & Biswas, D. (2014). A left-side bias? The influence of nutrition label placement on product evaluation. *ACR North American Advances*.

Roth, L. M., & Kroll, J. C. (2007). Risky business: Assessing risk preference explanations for gender differences in religiosity. *American Sociological Review*, 72(2), 205–220.

Rusconi, E., Kwan, B., Giordano, B. L., Umilta, C., & Butterworth, B. (2006). Spatial representation of pitch height: the SMARC effect. *Cognition*, 99(2), 113–129.

Russell, J. A. (1991). Culture and the categorization of emotions. *Psychological Bulletin*, 110(3), 426.

Russell, J. A. (2003). Core affect and the psychological construction of emotion. *Psychological Review*, 110(1), 145.

Sachdeva, S., Iliev, R., & Medin, D. L. (2009). Sinning saints and saintly sinners: The paradox of moral self-regulation. *Psychological Science*, 20(4), 523–528.

Sagristano, M. D., Trope, Y., & Liberman, N. (2002). Time-dependent gambling: Odds now, money later. *Journal of Experimental Psychology: General*, 131(3), 364.

Santana, E., & De Vega, M. (2011). Metaphors are embodied, and so are their literal counterparts. *Frontiers in Psychology*, 2, 90.

Sapir, E. (1929). A study in phonetic symbolism. *Journal of Experimental Psychology*, 12(3), 225.

Schjoedt, U., & Bulbulia, J. (2011). The need to believe in conflicting propositions. *Religion, Brain & Behavior*, 1(3), 236–239.

Schley, D. R., Lembregts, C., & Peters, E. (2017). The role of evaluation mode on the unit effect. *Journal of Consumer Psychology*, 27(2), 278–286.

Schloss, K. B., Strauss, E. D., & Palmer, S. E. (2013). Object color preferences. *Color Research & Application*, 38(6), 393–411.

Schlosser, A. E., Rikhi, R. R., & Dagogo-Jack, S. W. (2016). The ups and downs of visual orientation: The effects of diagonal orientation on product judgment. *Journal of Consumer Psychology*, 26(4), 496–509.

Schnall, S., Zadra, J. R., & Proffitt, D. R. (2010). Direct evidence for the economy of action: Glucose and the perception of geographical slant. *Perception*, 39(4), 464–482.

Schneider, I. K., Eerland, A., van Harreveld, F., Rotteveel, M., van der Pligt, J., Van der Stoep, N., & Zwaan, R. A. (2013). One way and the other: The bidirectional relationship between ambivalence and body movement. *Psychological Science*, 24(3), 319–325.

Scholl, B. J. (2001). Objects and attention: The state of the art. Cognition, 80(1–2), 1–46.

Scholl, B. J., & Leslie, A. M. (1999). Explaining the infant's object concept: Beyond the perception/cognition dichotomy. *What is Cognitive Science*, 26–73.

Schuldt, J. P., Konrath, S. H., & Schwarz, N. (2012). The right angle: Visual portrayal of products affects observers' impressions of owners. *Psychology & Marketing*, 29(10), 705–711.

Schwarz, N., & Clore, G. L. (1983). Mood, misattribution, and judgments of well-being: informative and directive functions of affective states. *Journal of Personality and Social Psychology*, 45(3), 513.

Ścigała, K., & Indurkhya, B. (2016). The influence of verticality metaphor on moral judgment and intuition. In *2016 7th IEEE International Conference on Cognitive Infocommunications* (CogInfoCom) (pp. 000205–000212). IEEE.

Sedgewick, J. R., Flath, M. E., & Elias, L. J. (2017). Presenting your best self (ie): The influence of gender on vertical orientation of selfies on tinder. *Frontiers in Psychology*, 8, 604.

Sell, A., Tooby, J., & Cosmides, L. (2009). Formidability and the logic of human anger. *Proceedings of the National Academy of Sciences*, 106(35), 15073–15078.

Semin, G. R., & Palma, T. A. (2014). Why the bride wears white: grounding gender with brightness. *Journal of Consumer Psychology*, 24(2), 217–225.

Shah, A. K., & Alter, A. L. (2014). Consuming experiential categories. *Journal of Consumer Research*, 41(4), 965–977.

Shariff, A., Norenzayan, A., & Henrich, J. (2010). The birth of high gods. *Evolution, Culture, and the Human Mind*, 119–136.

Shen, H., & Sengupta, J. (2012). If you can't grab it, it won't grab you: The effect of restricting the dominant hand on target evaluations. *Journal of Experimental Social Psychology*, 48(2), 525–529.

Shen, H., Wyer Jr, R. S., & Cai, F. (2012). The generalization of deliberative and automatic behavior: The role of procedural knowledge and affective reactions. *Journal of Experimental Social Psychology*, 48(4), 819–828.

Shen, L., & Urminsky, O. (2013). Making sense of nonsense: The visual salience of units determines sensitivity to magnitude. *Psychological Science*, 24(3), 297–304.

Shenhav, A., Rand, D. G., & Greene, J. D. (2012). Divine intuition: Cognitive style influences belief in God. *Journal of Experimental Psychology: General*, 141(3), 423.

Sherman, G. D., & Clore, G. L. (2009). The color of sin: White and black are perceptual symbols of moral purity and pollution. *Psychological Science*, 20(8), 1019–1025.

Shinners, E. (2009). Effects of the "what is beautiful is good" stereotype on perceived trustworthiness. *UW-L Journal of Undergraduate Research*, 12, 1–5.

Shoham, M., Moldovan, S., & Steinhart, Y. (2018). Mind the gap: How smaller numerical differences can increase product attractiveness. *Journal of Consumer Research*, 45(4), 761–774.

Sibley, C. G., & Bulbulia, J. (2012). Faith after an earthquake: A longitudinal study of religion and perceived health before and after the 2011 Christchurch New Zealand earthquake. *PloS One*, 7(12), e49648.

Sigall, H., & Ostrove, N. (1975). Beautiful but dangerous: effects of offender attractiveness and nature of the crime on juridic judgment. *Journal of Personality and Social Psychology*, 31(3), 410.

Simmons, W. K., Ramjee, V., Beauchamp, M. S., McRae, K., Martin, A., & Barsalou, L. W. (2007). A common neural substrate for perceiving and knowing about color. *Neuropsychologia*, 45(12), 2802–2810.

Simner, J., Ward, J., Lanz, M., Jansari, A., Noonan, K., Glover, L., & Oakley, D. A. (2005). Non-random associations of graphemes to colours in synaesthetic and non-synaesthetic populations. *Cognitive Neuropsychology*, 22(8), 1069–1085.

Slepian, M. L., & Galinsky, A. D. (2016). The voiced pronunciation of initial phonemes predicts the gender of names. *Journal of Personality and Social Psychology*, 110(4), 509.

Slepian, M. L., Camp, N. P., & Masicampo, E. J. (2015). Exploring the secrecy burden: Secrets, preoccupation, and perceptual judgments. *Journal of Experimental Psychology: General*, 144(2), e31.

Slepian, M. L., Masicampo, E. J., & Ambady, N. (2015). Cognition from on high and down low: Verticality and construal level. *Journal of Personality and Social Psychology*, 108(1), 1.

Slepian, M. L., Rule, N. O., & Ambady, N. (2012). Proprioception and person perception: Politicians and professors. *Personality and Social Psychology Bulletin*, 38(12), 1621–1628.

Slepian, M. L., Weisbuch, M., Rule, N. O., & Ambady, N. (2011). Tough and tender: Embodied categorization of gender. *Psychological Science*, 22(1), 26–28.

Small, D. A., & Loewenstein, G. (2003). Helping a victim or helping the victim: Altruism and identifiability. *Journal of Risk and Uncertainty*, 26(1), 5–16.

Smith, R. K., Newman, G. E., & Dhar, R. (2015). Closer to the creator: Temporal contagion explains the preference for earlier serial numbers. *Journal of Consumer Research*, 42(5), 653–668.

Snarey, J. (1996). The natural environment's impact upon religious ethics: A cross-cultural study. *Journal for the Scientific Study of Religion*, 85–96.

Song, H., Vonasch, A. J., Meier, B. P., & Bargh, J. A. (2012). Brighten up: Smiles facilitate perceptual judgment of facial lightness. *Journal of Experimental Social Psychology*, 48(1), 450–452.

Spector, F., & Maurer, D. (2011). The colors of the alphabet: Naturally-biased associations between shape and color. *Journal of Experimental Psychology: Human Perception and Performance*, 37(2), 484.

Spruyt, A., Hermans, D., Houwer, J. D., & Eelen, P. (2002). On the nature of the affective priming effect: Affective priming of naming responses. *Social Cognition*, 20(3), 227–256.

Stanfield, R. A., & Zwaan, R. A. (2001). The effect of implied orientation derived from verbal context on picture recognition. *Psychological Science*, 12(2), 153–156.

Stefanucci, J. K., Proffitt, D. R., Clore, G. L., & Parekh, N. (2008). Skating down a steeper slope: Fear influences the perception of geographical slant. *Perception*, 37(2), 321–323.

Strack, F., & Neumann, R. (2000). Furrowing the brow may undermine perceived fame: The role of facial feedback in judgments of celebrity. *Personality and Social Psychology Bulletin*, 26(7), 762–768.

Strahilevitz, M., & Myers, J. G. (1998). Donations to charity as purchase incentives: How well they work may depend on what you are trying to sell. *Journal of Consumer Research*, 24(4), 434–446.

Strauss, E. D., Schloss, K. B., & Palmer, S. E. (2013). Color preferences change after experience with liked/disliked colored objects. *Psychonomic Bulletin & Review*, 20(5), 935–943.

Sugovic, M., Turk, P., & Witt, J. K. (2016). Perceived distance and obesity: It's what you weigh, not what you think. *Acta Psychologica*, 165, 1–8.

Sun, Y., Wang, F., & Li, S. (2011). Higher height, higher ability: Judgment confidence as a function of spatial height perception. *PloS One*, 6(7), e22125.

Sundar, A., & Noseworthy, T. J. (2014). Place the logo high or low? Using conceptual metaphors of power in packaging design. *Journal of Marketing*, 78(5), 138–151.

Susskind, J. M., Lee, D. H., Cusi, A., Feiman, R., Grabski, W., & Anderson, A. K. (2008). Expressing fear enhances sensory acquisition. *Nature Neuroscience*, 11(7), 843.

Taylor, J. E. T., & Witt, J. K. (2010). When walls are no longer barriers: Perception of obstacle height in parkour. *Journal of Vision*, 10(7), 1017–1017.

Tenenbaum, H. R., & Leaper, C. (2003). Parent-child conversations about science: The socialization of gender inequities? *Developmental Psychology*, 39(1), 34.

Thaler, R. (1985). Mental accounting and consumer choice. *Marketing Science*, 4(3), 199–214.

Thomas, M., Simon, D. H., & Kadiyali, V. (2010). The price precision effect: Evidence from laboratory and market data. *Marketing Science*, 29(1), 175–190.

Todd, A. R., Hanko, K., Galinsky, A. D., & Mussweiler, T. (2011). When focusing on differences leads to similar perspectives. *Psychological Science*, 22(1), 134–141.

Todorov, A., Mandisodza, A. N., Goren, A., & Hall, C. C. (2005). Inferences of competence from faces predict election outcomes. *Science*, 308(5728), 1623–1626.

Todorov, A., Pakrashi, M., & Oosterhof, N. N. (2009). Evaluating faces on trustworthiness after minimal time exposure. *Social Cognition*, 27(6), 813–833.

Topolinski, S., Maschmann, I. T., Pecher, D., & Winkielman, P. (2014). Oral approach–avoidance: Affective consequences of muscular articulation dynamics. *Journal of Personality and Social Psychology*, 106(6), 885.

Treisman, A., & Schmidt, H. (1982). Illusory conjunctions in the perception of objects. *Cognitive Psychology*, 14(1), 107–141.

Tremoulet, P. D., & Feldman, J. (2000). Perception of animacy from the motion of a single object. *Perception*, 29(8), 943–951.

Trivers, R. L. (1971). The evolution of reciprocal altruism. *The Quarterly Review of Biology*, 46(1), 35–57.
Trope, Y., & Liberman, N. (2010). Construal-level theory of psychological distance. *Psychological Review*, 117(2), 440.
Tu, Y., & Soman, D. (2014). The categorization of time and its impact on task initiation. Journal of Consumer Research, 41(3), 810–822.
Turner, Y., & Hadas-Halpern, I. (2008, February). The effects of including a patient's photograph to the radiographic examination. In *Radiological Society of North America scientific assembly and annual meeting*. Oak Brook, Ill: Radiological Society of North America (Vol. 576).
Tversky, A., & Kahneman, D. (1973). Availability: A heuristic for judging frequency and probability. *Cognitive Psychology*, 5(2), 207–232.
Umilta, M. A., Berchio, C., Sestito, M., Freedberg, D., & Gallese, V. (2012). Abstract art and cortical motor activation: an EEG study. *Frontiers in Human Neuroscience*, 6, 311.
Vail, K. E., Rothschild, Z. K., Weise, D. R., Solomon, S., Pyszczynski, T., & Greenberg, J. (2010). A terror management analysis of the psychological functions of religion. *Personality and Social Psychology Review*, 14(1), 84–94.
Van Boven, L., & Ashworth, L. (2007). Looking forward, looking back: Anticipation is more evocative than retrospection. *Journal of Experimental Psychology: General*, 136(2), 289.
van de Ven, N., van Rijswijk, L., & Roy, M. M. (2011). The return trip effect: Why the return trip often seems to take less time. *Psychonomic Bulletin & Review*, 18(5), 827.
Van Dijck, J. P., Fias, W., & Andres, M. (2015). Selective interference of grasp and space representations with number magnitude and serial order processing. *Psychonomic Bulletin & Review*, 22(5), 1370–1376.
Van Kerckhove, A., & Pandelaere, M. (2018). Why Are You Swiping Right? The Impact of Product Orientation on Swiping Responses. *Journal of Consumer Research*, 45(3), 633–647.
Van Kerckhove, A., Geuens, M., & Vermeir, I. (2014). The floor is nearer than the sky: How looking up or down affects construal level. *Journal of Consumer Research*, 41(6), 1358–1371.
Van Quaquebeke, N., & Giessner, S. R. (2010). How embodied cognitions affect judgments: Height-related attribution bias in football foul calls. *Journal of Sport and Exercise Psychology*, 32(1), 3–22.
Van Rompay, T. J., De Vries, P. W., Bontekoe, F., & Tanja-Dijkstra, K. (2012). Embodied product perception: Effects of verticality cues in advertising and packaging design on consumer impressions and price expectations. *Psychology & Marketing*, 29(12), 919–928.
van Ulzen, N. R., Semin, G. R., Oudejans, R. R., & Beek, P. J. (2008). Affective stimulus properties influence size perception and the Ebbinghaus illusion. *Psychological Research*, 72(3), 304–310.
Veltkamp, M., Aarts, H., & Custers, R. (2008). Perception in the service of goal pursuit: Motivation to attain goals enhances the perceived size of goal-instrumental objects. *Social Cognition*, 26(6), 720–736.
Wadhwa, M., & Zhang, K. (2014). This number just feels right: The impact of roundedness of price numbers on product evaluations. *Journal of Consumer Research*, 41(5), 1172–1185.
Wagner, S., Winner, E., Cicchetti, D., & Gardner, H. (1981). "Metaphorical" Mapping in Human Infants. *Child Development*, 728–731.
Walker, P. (2015). Depicting visual motion in still images: forward leaning and a left to right bias for lateral movement. *Perception*, 44(2), 111–128.
Walker, P., & Walker, L. (2012). Size–brightness correspondence: Crosstalk and congruity among dimensions of connotative meaning. *Attention, Perception, & Psychophysics*, 74(6), 1226–1240.

Weger, U. W., Meier, B. P., Robinson, M. D., & Inhoff, A. W. (2007). Things are sounding up: Affective influences on auditory tone perception. *Psychonomic Bulletin & Review*, 14(3), 517–521.
Weinberg, S. (1979). Interview. Retrieved: https://www.youtube.com/watch?v=0gSomorLJQU
Whalen, D. H., & Levitt, A. G. (1995). The universality of intrinsic F0 of vowels. *Journal of Phonetics*, 23(3), 349–366.
Wilcox, K., Vallen, B., Block, L., & Fitzsimons, G. J. (2009). Vicarious goal fulfillment: When the mere presence of a healthy option leads to an ironically indulgent decision. *Journal of Consumer Research*, 36(3), 380–393.
Wilkie, J. E., & Bodenhausen, G. V. (2015). The numerology of gender: gendered perceptions of even and odd numbers. *Frontiers in Psychology*, 6, 810.
Wilkowski, B. M., Meier, B. P., Robinson, M. D., Carter, M. S., & Feltman, R. (2009). "Hotheaded" is more than an expression: The embodied representation of anger in terms of heat. *Emotion*, 9(4), 464.
Williams, L. E., & Bargh, J. A. (2008a). Experiencing physical warmth promotes interpersonal warmth. *Science*, 322(5901), 606–607.
Williams, L. E., & Bargh, J. A. (2008b). Keeping one's distance: The influence of spatial distance cues on affect and evaluation. *Psychological Science*, 19(3), 302–308.
Williams, L. E., Huang, J. Y., & Bargh, J. A. (2009). The scaffolded mind: Higher mental processes are grounded in early experience of the physical world. *European Journal of Social Psychology*, 39(7), 1257–1267.
Wilson-Mendenhall, C. D., Barrett, L. F., Simmons, W. K., & Barsalou, L. W. (2011). Grounding emotion in situated conceptualization. *Neuropsychologia*, 49(5), 1105–1127.
Wilson, G. D., & Barrett, P. T. (1987). Parental characteristics and partner choice: some evidence for Oedipal imprinting. *Journal of Biosocial Science*, 19(2), 157–161.
Wilson, R. K., & Eckel, C. C. (2006). Judging a book by its cover: Beauty and expectations in the trust game. *Political Research Quarterly*, 59(2), 189–202.
Winawer, J., Witthoft, N., Frank, M. C., Wu, L., Wade, A. R., & Boroditsky, L. (2007). Russian blues reveal effects of language on color discrimination. *Proceedings of the National Academy of Sciences*, 104(19), 7780–7785.
Windhager, S., Hutzler, F., Carbon, C. C., Oberzaucher, E., Schaefer, K., Thorstensen, T., . . . & Grammer, K. (2010). Laying eyes on headlights: Eye movements suggest facial features in cars. *Collegium Antropologicum*, 34(3), 1075–1080.
Winkielman, P., & Cacioppo, J. T. (2001). Mind at ease puts a smile on the face: psychophysiological evidence that processing facilitation elicits positive affect. *Journal of personality and social psychology*, 81(6), 989.
Winkler, I., Háden, G. P., Ladinig, O., Sziller, I., & Honing, H. (2009). Newborn infants detect the beat in music. *Proceedings of the National Academy of Sciences*, 106(7), 2468–2471.
Witt, J. K., & Proffitt, D. R. (2008). Action-specific influences on distance perception: a role for motor simulation. *Journal of Experimental Psychology: Human Perception and Performance*, 34(6), 1479.
Witt, J. K., Proffitt, D. R., & Epstein, W. (2005). Tool use affects perceived distance, but only when you intend to use it. *Journal of Experimental Psychology: Human Perception and Performance*, 31(5), 880.
Xie, J., Lu, Z., Wang, R., & Cai, Z. G. (2016). Remember hard but think softly: Metaphorical effects of hardness/softness on cognitive functions. *Frontiers in Psychology*, 7, 1343.
Xuan, B., Zhang, D., He, S., & Chen, X. (2007). Larger stimuli are judged to last longer. *Journal of Vision*, 7(10), 2–2.

Yamagishi, K. (1997). When a 12.86% mortality is more dangerous than 24.14%: Implications for risk communication. *Applied Cognitive Psychology: The Official Journal of the Society for Applied Research in Memory and Cognition*, 11(6), 495–506.

Yan, D. (2016). Numbers are gendered: The role of numerical precision. *Journal of Consumer Research*, 43(2), 303–316.

Yao, B., & Scheepers, C. (2011). Contextual modulation of reading rate for direct versus indirect speech quotations. *Cognition*, 121(3), 447–453.

Yao, B., Belin, P., & Scheepers, C. (2012). Brain 'talks over' boring quotes: Top-down activation of voice-selective areas while listening to monotonous direct speech quotations. *NeuroImage*, 60(3), 1832–1842.

Yaxley, R. H., & Zwaan, R. A. (2007). Simulating visibility during language comprehension. *Cognition*, 105(1), 229–236.

Zhang, C., Lakens, D., & IJsselsteijn, W. A. (2015). The illusion of nonmediation in telecommunication: Voice intensity biases distance judgments to a communication partner. *Acta Psychologica*, 157, 101–105.

Zhang, M., & Wang, J. (2009). Psychological distance asymmetry: The spatial dimension vs. other dimensions. *Journal of Consumer Psychology*, 19(3), 497–507.

Zhang, Y., & Risen, J. L. (2014). Embodied motivation: Using a goal systems framework to understand the preference for social and physical warmth. *Journal of Personality and Social Psychology*, 107(6), 965.

Zhao, M., & Xie, J. (2011). Effects of social and temporal distance on consumers' responses to peer recommendations. *Journal of Marketing Research*, 48(3), 486–496.

Zhao, M., Lee, L., & Soman, D. (2012). Crossing the virtual boundary: The effect of task-irrelevant environmental cues on task implementation. *Psychological Science*, 23(10), 1200–1207.

Zheng, X., Fehr, R., Tai, K., Narayanan, J., & Gelfand, M. J. (2015). The unburdening effects of forgiveness: Effects on slant perception and jumping height. *Social Psychological and Personality Science*, 6(4), 431–438.

Zhong, C. B., & Leonardelli, G. J. (2008). Cold and lonely: Does social exclusion literally feel cold? *Psychological Science*, 19(9), 838–842.

Zuckerman, M., Silberman, J., & Hall, J. A. (2013). The relation between intelligence and religiosity: A meta-analysis and some proposed explanations. *Personality and Social Psychology Review*, 17(4), 325–354.

Zuckerman, P. (2009). Atheism, secularity, and well-being: How the findings of social science counter negative stereotypes and assumptions. *Sociology Compass*, 3(6), 949–971.

Zwaan, R. A. (1996). Processing narrative time shifts. *Journal of Experimental Psychology: Learning, Memory, and Cognition*, 22(5), 1196.

Zwaan, R. A., Stanfield, R. A., & Yaxley, R. H. (2002). Language comprehenders mentally represent the shapes of objects. *Psychological Science*, 13(2), 168–171.

Printed in Great Britain
by Amazon